T0224855

FAUNA ENTOMOLOGICA SCANDINAVICA

Volume 1   1973

The Stratiomyioidea (Diptera) of Fennoscandia and Denmark

by

R. Rozkošný

SCANDINAVIAN SCIENCE PRESS LTD.

Gadstrup . Denmark

Copyright for the World
Scandinavian Science Press Ltd.

Edited by
Societas Entomologica Scandinavica

Editor of this volume
Leif Lyneborg

World list abbreviation
Fauna ent. scand.

Printed by
Vinderup Bogtrykkeri,
7830 Vinderup, Denmark

ISBN 87-87491-00-1

Contents

Introduction

The present study deals with the northern Solvidae and Stratiomyidae represented in the territories of Denmark and Fennoscandia, viz., Norway,
Sweden, Finland and the Soviet part of eastern Fennoscandia. Both families
apparently form a monophyletic group, generally called the superfamily
Stratiomyioidea, which is characterized by a broad conformity in the larval
morphology, presence of a compact prosternal bridge in the adults, absence
of the costa on the posterior margin of the wings, and characteristically
curved vein Cu1b connected with 1A before the wing-margin (Hennig, 1967).

In the area treated here, the Stratiomyioidea are represented by 19 genera
containing 50 species. The male terminalia are used extensively as taxonomic
characters for each species and for this reason deserve special attention.
Mature larvae and puparia, as far as they were available, are reexamined
and their chaetotaxy in particular is studied in great detail. This paper also
includes recent changes in nomenclature, the main information on diagnostic
characters and variability of adults and immature stages, distribution data
and biological notes. Keys have been provided for all adults and known larvae
at the generic and specific levels.

The distribution data discussed in the text and summarized in the appended
tables are based on the examination and revision of extensive material
deposited in the dipterological collections of Copenhagen, Bergen, Oslo, Lund,
Stockholm, Göteborg and Helsinki as well as in some private collections.
Most of the type specimens of the species described by J.C.Fabricius, C.F.
Fallén, J.W.Zetterstedt, R.C.Stæger and R.Frey have been studied.

Acknowledgements

For the kind loan of material from various institutions I am much indebted
to Mr. L. Lyneborg, Zoological Museum, Copenhagen; Mrs. Astrid Løken and
Prof. Dr. H. Kauri, Zoological Museum, Bergen; Dr. A. Lillehammer, Zoological
Museum, Oslo; Mr. H. Andersson, Zoological Institute, Lund; Mr. P. I. Persson,
Natural History Museum, Stockholm; Dr. H. W. Waldén, Natural History Museum,
Göteborg and Dr. W. Hackman, Zoological Museum, Helsinki. Mr. E. Torp
Pedersen, Jelling, Denmark, and Mr. T. R. Nielsen, Sandnes, Norway, kindly
provided specimens for study from their private collections.

In addition Mr. L. Lyneborg, Dr. A. Lillehammer, Mr. H. Andersson and Dr. W. Hackman located many collecting sites, interpreted old spellings of place names, and helped with their assignment to the various biological provinces. Some essential type and comparative material was kindly supplied by Mr. C. W. Berg, Hope Department of Entomology, Oxford; Prof. M. T. James, Washington State University, Pullman; Mr. K. G. V. Smith, British Museum (Natural History), London; Dr. H. Schumann, Museum für Naturkunde, Berlin and Mrs. E. P. Nartshuk, Zoological Institute, Leningrad.

Finally, I am very indebted to Mr. L. Lyneborg for his many-sided support during my research stay in Copenhagen as well as his helpful criticisms and suggestions for this study, and to Mrs. G. Lyneborg for making the figures of the general appearances and the final drawings of the adult characters.

Literature on the Stratiomyioidea
of Denmark and Fennoscandia

Owing to the relatively plentiful and mainly reliable literary records on the Stratiomyioidea of the area in question, particular attention has been paid to the relevant literary sources.

The first species of Stratiomyidae were described in the classical work by Linné (1758, 1767) and these descriptions are probably all based on material taken in Sweden (cf. Linné, 1761). In the extensive entomological work by Fabricius, 15 new species of Palaearctic Stratiomyidae are included, of which Stratiomys tigrina and S. viridula are described from the territory of Denmark (Fabricius, 1775). The first monograph on Swedish Stratiomyidae, also including the description of Oxycera pygmaea, was published by Fallén (1817, addenda 1826). Further new species were described in Zetterstedt's Insecta Lapponica (1838) and in his famous work Diptera Scandinaviae (Zetterstedt, 1842-1859). A paper by Stæger (1844) represents an important contribution to the taxonomy of Oxycera. Some Stratiomyidae new to Sweden were added by Wahlberg (1854), including the new species Sargus rufipes. We can thus state that nearly half of the 50 recorded species of Stratiomyioidea were described by native authors.

Faunal data are scattered through numerous papers of rather variable value. Relatively little information is available on Norwegian species (Siebke, 1863, 1877; Schøyen, 1889; Bidenkap, 1900; Strand, 1903; Storm, 1907). The Danish species were comprehensively worked out in the first part of the excellent work by Lundbeck (1907), and the faunal records were completed later by Lundbeck himself (1914, 1919), Jørgensen (1917) and Lyneborg (1960, 1965). A brief survey of the Swedish Stratiomyidae was published by Wahlgren (1907) and the discovery of Solva marginata was reported by the same author (Wahlgren, 1921). Many faunal records are to be found in numerous entomological papers by Wallengren (1866, 1870), Jansson (1922, 1925, 1935), Lindroth (1943), Lundblad (1950, 1954, 1955) and Ardö (1957). Ringdahl's dipterological studies (1914, 1917, 1931, 1935, 1937, 1939, 1941, 1947, 1952, 1954, 1958, 1959a, 1959b, 1960) are especially worthy of mention. Papers by Hanson (1942) and Andersson (1962, 1971a, 1971b) are also important from the taxonomic point-of-view.

In eastern Fennoscandia Stratiomyioidea were first listed by Bonsdorff (1861) and later by Frey (1911, 1941). In addition to Frey's faunal papers (e.g. 1918, 1947), his contribution on Oxycera freyi Lind. (Frey, 1953) is particularly valuable, and some notes on Solvidae within the framework of a larger study (Frey, 1960) are also of certain taxonomic importance.

Furthermore, some distribution data and taxonomic remarks are to be found in various monographs or special studies by authors outside Denmark and Fennoscandia (e.g. Pleske, 1924; Lindner, 1935-38; Stackelberg, 1954; Dušek & Rozkošný, 1968).

Characteristics of Stratiomyioidea

The general appearance of the superfamily is very diverse, and Scandinavian species of this group include forms from about 2.0 mm to 18.0 mm in length. Some of them are rather slender, whilst others are stout or conspicuously flattened. The coloration is often striking, being dark with yellow, white or greenish pattern, or sometimes blue or green with metallic reflections.

The head is mostly semiglobular, transversely broadened in some species, with the face sometimes more or less protuberant or even elongated into a conical rostrum. Antennae with the 3rd segment annulated, consisting of 4-8 larger flagellomeres, elongate or spindle-shaped with a more or less distinct apical style, or oval to disc-shaped with terminal or subterminal arista. Females always dichoptic, males mostly holoptic. Palpi two-segmented or rudimentary.

Thorax usually almost rectangular. Scutellum with 2-8 marginal spines or unarmed. Costal vein on wings reaching somewhat beyond the tip of veins R5 (Stratiomyidae), or M1 or M2 (Solvidae), absent on the posterior margin of

wing. R4 present, or absent in some cases. Three or four M-veins arising
from the discal cell and consequently a cross-vein between the discal cell and
M4 developed or absent. M-veins often very fine, shortened or even rudimentary. M3 and M4 connected before wing margin in the Solvidae, as are veins
Cu1b and 1A in all Stratiomyioidea. Alula always well-developed, squamae
small or partly reduced. Legs simple, only exceptionally spurred (Solvidae
and Exodontha), with pad-like pulvilli and empodium.

Abdomen consisting of 5 to 7 clearly visible main segments, apical segments
telescoped into the apical part of abdomen. The shape of the abdomen is rather
variable, elongate, oval or large and flattened, sometimes short and almost
circular. The male genitalia are formed by the ninth and the following segments,
though the preceding segments may also be partly reduced or of rather unusual
structure. Tergite 9 (and probably also 10) is developed as a variable dorsal
plate - epandrium (e) bearing posteriorly an anal cone or proctiger (pr) with
a pair of cerci (c). If the proctiger is in fact identical with abdominal segment
11, then its single dorsal plate represents the epiproct (ep) and a pair of ventrolateral plates corresponds to the paraprocts (pp) of some more primitive
groups of insects.

The ventral part of the hypopygium is usually compact, consisting of fused
hypandrium (= sternite 9) and basal parts of the gonopodes (basistyles). This
structure is called the synsternite (s) by some authors. Its posterior margin
may be prolonged into a simple or bilobed, high or low medial process (mp).
The apical parts of the original gonopodes, the dististyles (dst), are free and
one-segmented. The male hypopygium may be complicated by posterolateral
projections of the epandrium (surstyles, ss), and by longer or shorter inner
or outer projections of the synsternite, and even by so-called ventral lobes
(vl) of the synsternite.

The aedeagal complex is composed of the aedeagus (ae) and the lateral parameres (p) which are usually fused with the aedeagus in its basal part, where a
flat aedeagal apodeme (ap) is distinct in some species. The aedeagal complex
is attached to special, usually strip-like, dorsal processes of the synsternite
(dp). These dorsal processes are sometimes connected basally, in which case
they form a more or less distinct dorsal bridge which may cover the aedeagal
complex dorsally. In the Pachygasterinae there appears a tendency for the
development of a large hyaline tube for the apical part of aedeagal complex,
consisting of the dorsal bridge and a similar elongated posteromedian part of
synsternite.

10

Immature stages,
biology and distribution notes

The eggs are elongate-oval, spindle-shaped, and often moderately flattened. No special structure of their external surface was noticed. In general they are relatively small, e.g. eggs of Clitellaria ephippium are about 1.2-1.5 mm in length. Aquatic species in particular lay their eggs in a conspicuous oval mass (about 13 mm long in Stratiomys), stuck together by a cementing substance (Lundbeck, 1907).

Larvae of Stratiomyioidea may be divided in two large groups according to their general appearance. Terrestrial larvae are essentially elongate-oval, rather broad and depressed, with a rounded anal segment. Aquatic larvae are also mostly flattened but are characterized by a more or less tapering posterior end, which bears an apical coronet of pinnate float-hairs, or by two strikingly prominent posterior lobes on the anal segment. The cuticle of all larvae is rough, reticulate, strongly sclerotised and encrusted with calcium carbonate.

The larval body consists of the head, 3 thoracic and 8 abdominal segments. The relatively small and regularly slender head is permanently retracted for about one-half of its length into the thorax. The two-segmented antennae (A) are usually inconspicuous, and are located in the anterolateral angles of the head or somewhat proximally. The mandibular-maxillary complex (Mm) is visible on each side of the tapered labrum. The eye prominences (Ep) are situated on the dorsolateral margin of head.

The first thoracic segment bears large anterior spiracles (As) laterally which vary in size and shape. The abdominal segments are similar to the thoracic segments except for the anal segment. Most aquatic larvae possess small and inconspicuous rudimentary spiracles (Sp) on the metathorax and first 6-7 abdominal segments. Some larvae also have more or less developed projections resembling papillae at the anterior corners of some abdominal segments, the function of which is rather problematic. Some authors (e. g. Austen, 1899) interpreted them as the remnants of once functional spiracles. On the other hand, these projections are very prominent in some Stratiomys larvae on segments where circular rudimentary spiracles are also distinct. The posterior spiracular opening (Ps) consists of a transverse cleft leading into the spiracular chamber and may be located dorsally or horizontally at the apex of the anal segment. Larvae of Stratiomyioidea were therefore originally peripneustic, but are functionally mostly amphipneustic or even metapneustic (Hennig, 1952).

11

On the mid-ventral line of abdominal segment 6 a so-called sternal patch (Spa) may be present. It is well-developed in most terrestrial larvae as a round or elongate-oval field with a different colour and often without any cuticular structure. The sternal patch is almost indistinct in larvae of Beris, and absent in larva of Clitellaria ephippium and in all aquatic larvae. The anus consists of a longitudinal ventral slit in the middle of anal segment.

The general scheme of chaetotaxy is as follows: On the head there are usually 4 pairs of dorsal setae, viz., 2 labral (Lb1, Lb2) and 2 clypeofrontal (Cf1, Cf2), 1 pair of dorsolateral setae (Dl) behind the eye prominence, 1 pair of lateral setae (L) between antennae and eyes, 3 pairs of ventrolateral setae (Vl1, Vl2, Vl3) which are usually preantennal, subantennal and postocular in position, and 3 pairs of ventral setae (V1, V2, V3).

All 3 thoracic segments bear 3 pairs of dorsal setae (D1, D2, D3), 1 pair of dorsolateral setae (Dl), 1 pair of ventrolateral setae (Vl) and two pairs of ventral setae (V1, V2), the inner one of which is simple and the outer one (V2) branched or consisting of 2-3 smaller setae and which is then known as thoracic leg group. In addition, thoracic segment 1 also has 2-3 pairs of anterodorsal setae (Ad1, Ad2, Ad3).

Abdominal segments 1-7 have very similar bristles, though the setae often become longer and stronger caudally. There are three pairs of dorsal setae situated as on the thoracic segments. In addition to normal Dl and Vl setae on the lateral wall of each segment, there are also 1-2 lateral setae (L1, L2) present. Finally, there are 3 pairs of ventral setae that are usually arranged in a transverse row or in a semicircular row. The last abdominal segment (= anal segment) shows a surprising unity in chaetotaxy in different species considering its variable form in terrestrial and aquatic larvae. Only 1 pair of dorsal setae (D) is more or less developed. 2 pairs of lateral setae (L1, L2) are often among the longest setae of the anal segment, as is 1 pair of subapical setae (Sa). The pair of apical setae (Ap) is usually relatively short. The location and number of the ventral setae is somewhat variable in different species, but usually 5 pairs of ventral setae are placed on the preanal wall and along or behind the anal slit.

As regards biology, there is a considerable diversity in habitat of the immature stages of Stratiomyioidea. Females of terrestrial species oviposit on earth, manure, decaying vegetable matter, and aquatic species on aquatic plants or damp moss. The actual time required for hatching varies from 5 days to 3 weeks (McFadden, 1967).

About half of the species treated here lives as terrestrial forms in the larval stage, whilst the other half occurs in various aquatic or at least semiaquatic

situations. The feeding habits of all larvae may be described as micropanto-
phagous, partly also as coprophagous or saprophagous. Larvae of Solvidae and
Pachygasterinae live under bark of both deciduous and coniferous trees, but a
certain specialisation appears in some species. Larvae of Beridinae are generally considered to be terrestrial and have actually been found in earth, under
stones and in decaying organic matter, but some of them also seem to be associated with semiaquatic situations. Sarginae have larvae living coprophagous-
ly and saprophagously. They are frequent in compost, garden refuse, manure,
cow dung, etc. The larva of Clitellaria ephippium, which occurs regularly in
the nests of ants, also belongs to the terrestrial forms.

The aquatic larvae belong to the subfamilies Stratiomyinae and Clitellariinae.
Whilst the larvae of Stratiomys, Odontomyia and Oplodontha prefer the litoral
zone of standing water, marshes and pools, many larvae of Oxycera spp. live
in moss cushions or among algae on stones and rocks in running water, springs
and streams, as members of the hygropetric fauna.

Puparia are formed from the last larval skin as a protection for the pupa
(pupa exarata), and are mostly found in the same habitat as the larva. Aquatic
larvae sometimes leave their environment before pupation, but on the other
hand their puparia sometimes float on the surface of the water. The adults
escape through a T-shaped longitudinal slit along the anterodorsal part of the
puparium.

Adults are generally found on waterside vegetation, in particular flowering
Daucaceae and Asteraceae, but also on ground-vegetation, the leaves of shrubs
and trees, etc. Males of some species (Nemotelus, Pachygasterinae) are reported to indulge in dancing flights or even to hover. Some species of Sarginae are
likely under certain conditions to transfer pathogenic microflora, and therefore
have a certain hygienic importance owing to their occurrence in dung and manure on the one hand and their evident synanthropic tendencies on the other hand.

The Stratiomyioidea are known to occur in all the main zoogeographical regions; nevertheless the centre of distribution of most groups seems to lie in
tropical and subtropical zones. Consequently, only about 150 species are found
in Europe, 50 of which also occur in the northern area treated here. Though
several species display a boreal type of distribution (e.g. Sargus rufipes, Oxycera freyi), only a few Beridinae, Sarginae and Nemotelus nigrinus reach to
the far north of Scandinavia. Apparently no Stratiomyioidea were collected on
the Faroes or on Iceland.

13

Key to families of Stratiomyioidea

Adults

1   Costa reaching the tip of vein M1 or M2, cell M3 closed
and stalked (Fig. 1); middle and hind tibiae spurred (Figs.
3-5); membranous area on the dorsal surface of abdominal segment 1 more or less distinct .... Solvidae

-    Costa reaching the tip of vein R5, cell M3 absent or open
(Fig. 33); tibiae without any spurs, or exceptionally with
a single spur on middle tibiae; membranous area on the
dorsal surface of abdominal segment 1 absent .... Stratiomyidae

Larvae

1   Anal slit bordered anteriorly by a transverse row of strong
posteriorly directed teeth (Figs. 12-13, 16); first two
thoracic segments usually with a smooth dorsal field (Figs.
8-10) .... Solvidae

-    Anal slit not bordered anteriorly by teeth (Figs. 81, 121,
192); all thoracic segments with a normal reticulate structure on dorsum .... Stratiomyidae

Family Solvidae (= Xylomyidae)

In comparison with the Stratiomyidae, the species of the Solvidae have the
eyes dichoptic in both sexes, third antennal segment consisting of 8 flagellomeres and sharply pointed at apex, scutellum always unarmed and cell M3
closed. The middle and hind tibiae bear a pair of spurs. The male genitalia
are rather complicated, and of specific value. The epiproct is mostly reduced,
the parameres not being fused basally with the aedeagus.

The larvae usually have a smooth shining area without any cuticular structure
on the middle of thoracic segments 1 and 2. The anal slit is bordered anteriorly by a transverse row of teeth. Body shape, cuticular structure, mouth parts
and general scheme of chaetotaxy resemble terrestrial larvae of Stratiomyidae.

The puparium is quite similar to the mature larva, without special puparial
characters, but the escape of the adults is original. The pupa partly emerges
from a normal T-shaped fissure on the dorsal surface of the anterior segments,
so that after emergence the empty pupal skin remains wedged by its posterior
extremity in the fissure or it may even be some distance away from the puparium.

Larvae live under the bark of trees and in rotting logs, often together with arboreal larvae of Stratiomyidae.

The taxonomy of Solvidae at the generic level has recently been discussed in detail. Hennig (1967) pointed out that Xylomya varia (Meig.), as the type-species of Xylomya, is much more closely related to Solva marginata (Meig.) than to the other species of "Xylomya". According to Nagatomi & Tanaka (1971), all Europe-

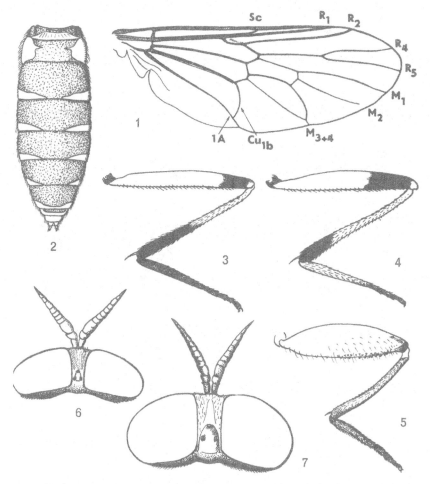

Figs.1-7. Morphology of adult Solvidae. - 1: Wing of S.marginata (Meig.); 2: Abdomen of female holotype of S.interrupta Pleske; 3: Hind leg of female of same; 4: Hind leg of male of S.maculata (Meig.); 5: Hind leg of female of S. marginata (Meig.); 6: Head of female of same; 7: Head of male of S. maculata (Meig.).

an species of Solvidae therefore only belong to the genus <u>Solva</u> which may be divided into two subgenera - <u>Solva</u> s.str. and <u>Macroceromys</u> Bigot (="Xylomya").

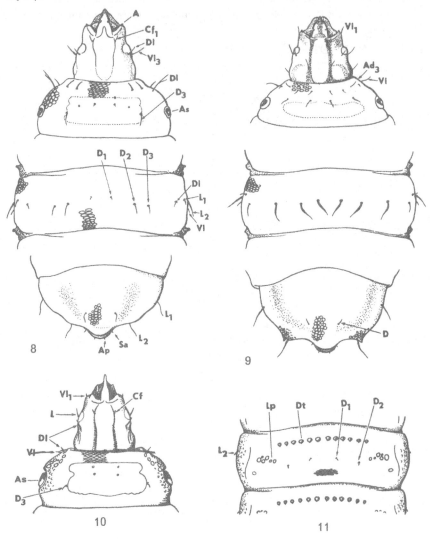

Figs. 8-11. Morphology of <u>Solva</u>-larvae. - 8: Anterior end, abdominal segment 3 and anal segment of <u>S.interrupta</u> Pleske; 9: Anterior end, abdominal segment 3 and anal segment of <u>S.maculata</u> (Meig.); 10: Head and thoracic segment 1 of <u>S.marginata</u> (Meig.); 11: Abdominal segment of same. Abbreviations: see p.132.

Genus <u>Solva</u> Walker, 1859

<u>Solva</u> Walker, 1859, Proc.Linn.Soc.Lond., 4:98.
<u>Xylomya</u> Rondani, 1861, Dipt.Ital.Prodr., 4:11.
Type-species: <u>Solva inamoena</u> Walker, 1859.

The main diagnostic features are given in the family diagnosis. Four European species; <u>varia</u> (Meig.) occurs only in England, C. and S. Europe.

Key to species of <u>Solva</u>

Adults

1   Hind femora distinctly thickened, with tubercles or spin-
    ules on ventral surface (Fig.5) .... 3. <u>marginata</u> (Meig.)
-   Hind femora not noticeably thicker than two anterior
    pairs, without ventral tubercles or spinules (Figs.3-4) 2
2 (1) Coxae yellow, hind metatarsi black (Fig.3)...... 1. <u>interrupta</u> Pleske
-   Coxae black, hind metatarsi yellow (Fig.4)....... 2. <u>maculata</u> (Meig.)

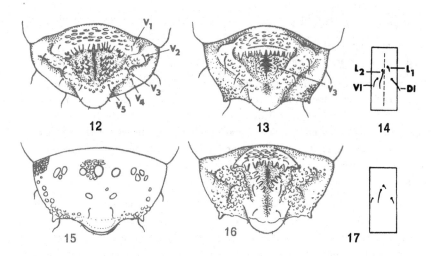

Figs.12-17. Morphology of <u>Solva</u>-larvae. - 12: Anal segment in ventral view of S.interrupta Pleske; 13: Anal segment in ventral view of S.maculata (Meig.); 14: Scheme of setae on lateral wall of abdominal segment 3 of same; 15: Anal segment in dorsal view of S.marginata (Meig.); 16: Anal segment in ventral view of same; 17: Scheme of setae on lateral wall of abdominal segment 3 of same. Abbreviations: see p.132.

Larvae

1   Prothoracic segment with a distinct incision in front of
       anterior spiracle, giving a cleft appearance (Fig. 10);
       dorsal tubercles on abdominal segments and larger cut-
       icular plates on anal segment present (Figs. 11, 15).....
       .... 3. marginata (Meig.)
-      Prothoracic segment without a cleft appearance in front
       of anterior spiracle; dorsal tubercles on abdominal seg-
       ments and larger cuticular plates on anal segment absent
       (Figs. 8-9).. 2
2 (1) Middle pair of dorsal setae (D1) reduced, very small;
       posterolateral angles of anal segment rounded (Fig. 8)..
       .... 1. interrupta Pleske
-      All dorsal setae almost equal and rather strong; postero-
       lateral angles of anal segment prominent, mostly pointed
       (Fig. 9) .... 2. maculata (Meig.)

1. SOLVA INTERRUPTA Pleske, 1926
   Figs. 2-3, 8, 12, 19-23.

Solva interrupta Pleske, 1926: 174.
Xylomya maculata var. sahlbergi Frey, 1960: 5.

Slender and relatively long species. All coxae yellow; darkening on hind femora narrower, and that on hind tibiae larger, than in maculata; hind metatarsi
entirely black. Yellow pleural markings smaller, and yellow hind-margin of
abdominal segments conspicuously narrowed or even interrupted medially. -
Male genitalia: Epandrium with rather short and wide surstyles, synsternite
with large ventral lobes, aedeagus transversely ridged in basal part. - Length:
body 10.0-13.0 mm, wing 9.5-11.0 mm.

   Larva with setae rather reduced. Only D1 and V13 setae more distinct on
head, as well as 1 pair of short Cf setae. Smooth central area on thoracic segment 1 large and well-defined, two pairs of Ad setae, longer and mainly
branched D1 setae, and of the dorsal setae mainly D3 distinct on this segment.
Abdominal segments with a transverse row of 6 inconspicuous dorsal setae,
the middle pair of which is very small. Posterolateral angles of anal segment
rounded. 9-11 longer cylindrical and blunt teeth in a transverse row in front
of anal slit, and two lateral groups, each with 9-14 blunt teeth alongsides anal

18

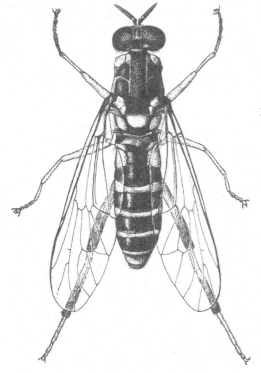

Fig. 18.

Male of Solva maculata (Meig.).

slit. - Length 14.0-16.5 mm, maximum width 3.2-4.8 mm (3 puparia in ZMH, 2 puparia in ZIL).

At present the species is known only from Finland, the Soviet part of eastern Fennoscandia and the region of Leningrad, as a vicariant for maculata. The known localities in Fennoscandia (Yläne, Orivesi, Kivinebb, Haapanava and Hattula) were all discovered by Sahlberg (1839), except for Hattula, and he also reared larvae and puparia from decaying aspen (Populus tremula). Part of this material was also mentioned by Zetterstedt (1842) under the name of maculata. - June. - Larvae under bark of aspen trunks, adults on trees.

Note: The relatively well-preserved female holotype of interrupta (only 1 antenna and 1 wing missing) is located in the Zoological Institute, Leningrad. It is identical with the male holotype of Frey's species from ZMH, bearing a large white label "Xylomyia ssp. Sahlbergi Frey, det. R. Frey" and a further red label with "Holotypus".

The larva described and figured by Krivosheina (1965), characterized by transverse rows of dorsal tubercles on the abdominal segments, seems to be more related to Solva marginata. It can hardly be conspecific with true interrupta.

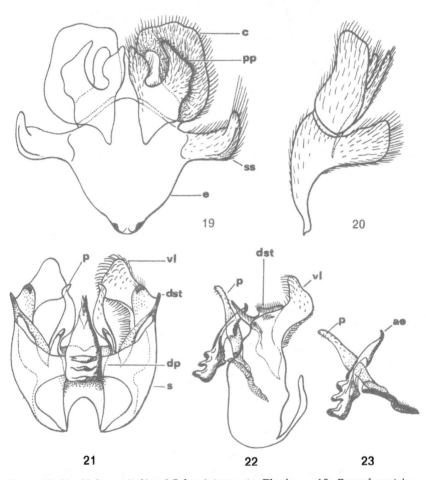

Figs. 19-23. Male genitalia of Solva interrupta Pleske. - 19: Dorsal part in ventral view; 20: Dorsal part in lateral view; 21: Ventral part in dorsal view; 22: Ventral part in lateral view; 23: Aedeagal complex in lateral view. Abbreviations: see p.132.

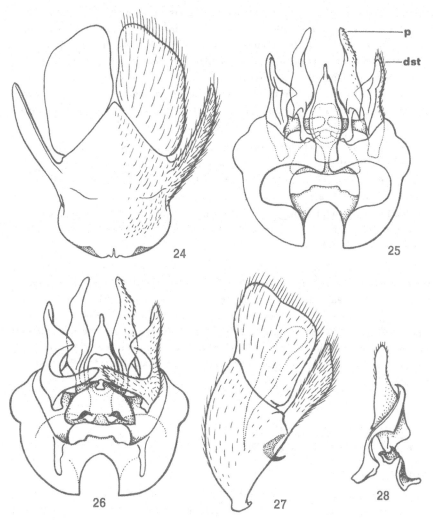

Figs. 24-28. Male genitalia of Solva maculata (Meig.). - 24: Dorsal part in dorsal view; 25: Ventral part in dorsal view; 26: Ventral part in ventral view; 27: Dorsal part in lateral view; 28: Aedeagal complex in lateral view. Abbreviations: see p. 132.

2. SOLVA MACULATA (Meigen, 1804)

Figs. 4, 7, 9, 13-14, 18, 24-28.

Xylophagus maculatus Meigen, 1804: 154.

Shining black and yellowish spotted species. Legs predominantly yellow, but all coxae black as well as apical third of hind femora and tibiae and also last 3-4 tarsal segments. Yellow pleural spot extending over upper mesopleura onto part of sterno- and pteropleura. Yellow hind-margins of abdominal segments mostly stripe-like, not extremely narrowed in the middle. - Male genitalia: Surstyles of epandrium rather slender and long, paraprocts simple. Dististyles large, ventral lobes of synsternite absent, aedeagus smooth in its basal part. - Length: body 8.0-12.5 mm, wing 7.5-12.0 mm.

Larva with well-developed setae on head as well as on body segments. Smooth central area on thoracic segment 1 usually rather narrow. 3 pairs of Ad setae and 2 pairs of D setae distinct on this segment. Abdominal segments with 3 pairs of D setae strong and equal. Posterolateral angles of anal segment pointed. 9-11 conical and mainly pointed teeth in front of anal slit, and laterally some similar teeth in irregular longitudinal rows. - Length: 15.0 mm, maximum width: 4.2 mm. (One larval skin from Central Europe).

Apparently absent from Fennoscandia except for S.Sweden where it was found in some localities in Scania (Östared, Lindholmen, Svedala). In Denmark known to occur in NE Zealand (Copenhagen, Dyrehaven, Frederiksdal, Ørholm) and Lolland (Maribo). - Throughout the greater part of Europe, the northernmost limits in England, Denmark and S.Sweden. A record of its occurrence in Siberia needs verification. - June-July. - Larva in rotten deciduous trees (Quercus, Fagus, Populus, Ulmus, Sorbus) and adults on trees and trunks in the same area.

3. SOLVA MARGINATA (Meigen, 1820)
   Figs.1, 5-6, 10-11, 15-17, 29-32.

Xylophagus marginatus Meigen, 1820: 15.

A medium-sized species, with rather short strong antennae, a transverse head, and black, punctate mesonotum without yellow markings. Antennae dark brown, the inner basal part often yellowish. Legs yellow with coxae and tips of hind femora black, and often also tips of tibiae and usually tips of tarsi. Hind femora conspicuously swollen, bearing minute blackish tubercles below. Male genitalia: Epandrium almost oval, without surstyles, cerci small. Synsternite without ventral lobes, dististyles large. Aedeagus compact and relatively long. - Length: body 5.0-8.0 mm, wing 5.5-8.0 mm.

The larva was inadequately described and figured by Dufour (1847), Séguy

(1926), Lindner (1936) and Brindle (1961), but detailed descriptions have been published by Dušek & Rozkošný (1963) and Krivosheina (1965). The brown to ochre-coloured larva has the usual net-like cuticular structure and distinct smooth areas in the middle of thoracic segments 1 and 2. Setae mostly very reduced. Only one pair of dorsal setae more developed on the body segments,

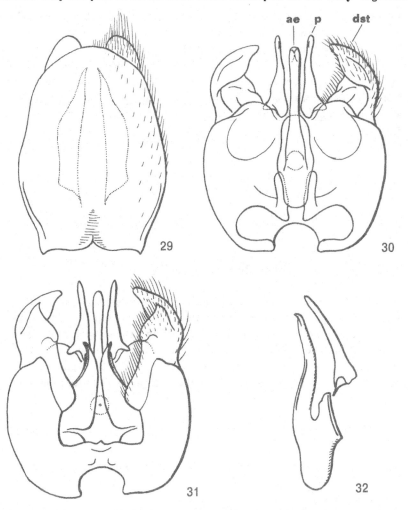

Figs. 29-32. Male genitalia of Solva marginata (Meig.). - 29: Dorsal part in dorsal view; 30: Ventral part in dorsal view; 31: Ventral part in ventral view; 32: Aedeagal complex in lateral view. Abbreviations: see p. 132.

at most 1/4 length of segment. A sharp lateral incision immediately in front of anterior spiracles. Each abdominal segment with a row of conical tubercles dorsally (6 to 12) and ventrally (14 to 19). Anal segment with a transverse row of teeth in front of anal slit and two lateral groups of teeth. - Length: 11.0-13.0 mm, maximum width: 2.5-2.8 mm. (Extensive material from Central Europe).

Very rare in N. Europe. Only one female has been taken in Sweden near Falsterbo, 6. VII. 1920 (Wahlgren), and a second has been reared from a larva found in Denmark, Jægerspris, VI-VII. 1943 (Anthon) (cf. also Wahlgren, 1921, and Lyneborg, 1960). - Mainly in S. and C. Europe including England, according to rather scattered distribution data. - May-August. - Larvae under bark and in rotten deciduous trees (Populus alba, P. tremula, P. nigra, Carpinus betulus, Robinia pseudoacacia, Aesculus, Salix). Adults mostly on trees, trunks and stumps.

Family Stratiomyidae

Eyes usually holoptic in males and dichoptic in females, scutellum with spines or unarmed. The number of flagellomeres of the third antennal segment variable in individual subfamilies. The costa reaches somewhat beyond the tip of R5, and cell M3 is always open. All the tibiae lack spurs, except Exodontha. The male genitalia are mainly rather simple. The epiproct is usually well-developed, and the parameres more or less fused with aedeagus in its basal part.

The larvae lack a smooth shining area on thoracic segments 1 and 2. No teeth are developed in front of the anal slit, though some species have the anal slit bordered by fine teeth laterally. The puparium is mostly quite similar to the mature larva, but special pupal spiracles are distinct in the puparia of some species. The true pupal skin remains completely in the puparium after emergence of the adult.

Larvae occur in various aquatic, semiaquatic and terrestrial situations. In Europe there are more than 150 species in 29 genera, and in the area treated here 47 species in 17 genera.

24

Key to subfamilies and genera of Stratiomyidae

Adults

1 Scutellum with at least two pairs of marginal spines (Fig. 35); abdomen consisting of seven main segments (except for Exodontha (Fig. 34) with five segments but spurred middle tibiae) (Beridinae) 2

- Scutellum either with a single pair of terminal spines or unspined; abdomen consisting of five main segments; tibiae without any spurs 3

2 (1) Palpi reduced; seven abdominal segments visible; body relatively slender (Fig. 50) .... Beris Latr. (p. 31)

- Palpi well-developed; only five main abdominal segments visible (Fig. 34); body robust .... Exodontha Rond. (p. 46)

3 (1) Cross-vein between the discal cell and M4 distinct (Fig. 33) 4

- Cross-vein between the discal cell and M4 absent, and M4 arising directly from the discal cell or absent (Figs. 39-41) 9

4 (3) Antennae relatively long; third segment spindle-shaped, without arista (Figs. 37-38) (Stratiomyinae) 7

- Antennae mainly short; third segment rounded or oval, with subterminal arista (Fig. 36) (Sarginae) 5

5 (4) Eyes with dense long hairs (Fig. 147) ...... Chloromyia Duncan (p. 64)

- Eyes bare or at most very sparsely haired 6

6 (5) Anal cell broad; eyes touching in males; M-veins fine, partly reduced .... Microchrysa Loew (p. 48)

- Anal cell relatively narrow; eyes distinctly separated in both sexes (Figs. 143-146); M-veins well-developed (Fig. 33) .... Sargus Fabr. (p. 55)

7 (4) First antennal segment 4 to 6 times as long as the second (Fig. 37) .... Stratiomys Geoffr. (p. 67)

- First antennal segment at most twice as long as the second (Fig. 38) 8

8 (7) Vein Rs unbranched; discal cell very small... Oplodontha Rond. (p. 87)

- Vein Rs distinctly branched; discal cell larger (Fig. 217) .... Odontomyia Meig. (p. 76)

9 (3)   4 M-veins always present (Fig.39); third antennal segment mostly spindle-shaped and with an apical style, rarely with subterminal arista (Clitellariinae) 10

-      Only 3 M-veins present (Figs.40-41); third antennal segment disc-shaped with subterminal arista, or oval with
terminal arista (Pachygasterinae) 12

10 (9)   Face cone-like, produced into a rostrum (Figs.284-299);
scutellum without spines .... Nemotelus Geoffr. (p.89)

-      Face not produced into a rostrum; scutellum with two spines.......11

11 (10) Thorax with a pair of strong spines in front of wing-base
.... Clitellaria Meig. (p.99)

-      Thorax without spines in front of wing-base.... Oxycera Meig. (p.100)

12 (9)   Third antennal segment disc-shaped, mostly wider than long....... 13

-      Third antennal segment oval, longer than wide (Figs.448,
450) .... Berkshiria Johnson (p.128)

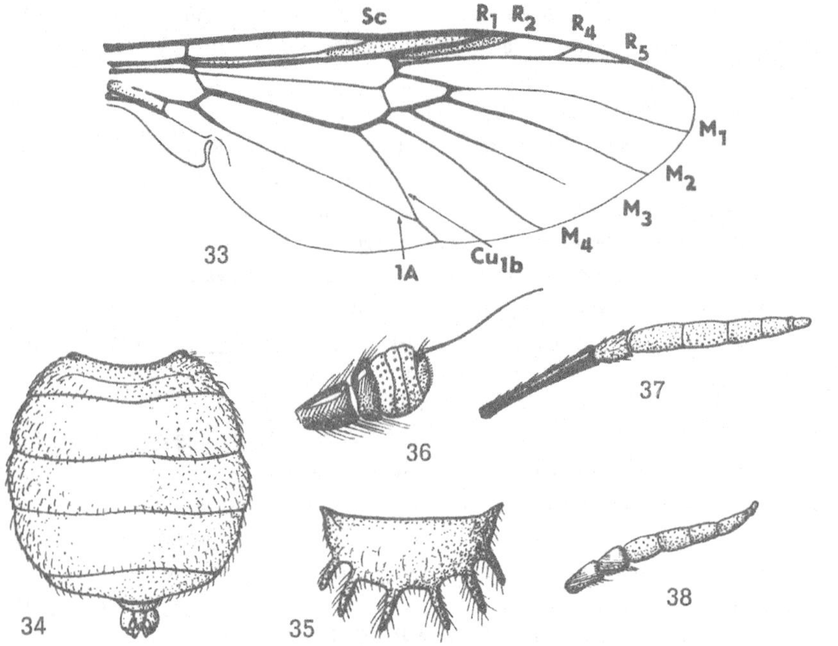

Figs. 33-38. Morphology of adult Stratiomyidae. - 33: Wing of Sargus iridatus
(Scop.); 34: Male abdomen of Exodontha dubia (Zett.); 35: Female scutellum
of Beris clavipes (L.); 36: Female antenna of Sargus iridatus (Scop.); 37: Male
antenna of Stratiomys chamaeleon (L.); 38: Female antenna of Odontomyia ornata (Meig.).

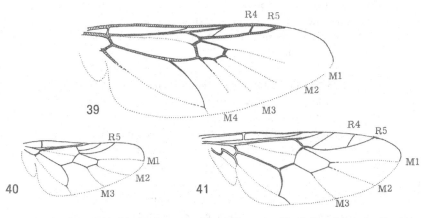

Figs. 39-41. Wings of: 39: Oxycera leonina (Panz.); 40: Zabrachia minutissima (Zett.); 41: Eupachygaster tarsalis (Zett.).

Larvae

1   Setae on body-segment surrounded by bristle-like hairs
        which form more or less distinct tufts (Fig. 75); anal
        segment rounded and fringed by fine setae (Fig. 76)
        (Beridinae) .... Beris Latr. (p. 31)

-       Setae on body-segments not surrounded by bristle-like
        hairs, though sometimes pubescent or plumate; anal seg-
        ment without marginal fringe 2

2 (1) Anal segment usually rounded posteriorly (Fig. 120, 148, 300, 303, 397), seldom bilobed, but never with a coronet of pinnate float-hairs at apex 3

\- Anal segment usually oblong, sometimes even elongated like a tube, with a coronet of pinnate float-hairs at apex (Figs. 192, 232, 344) 13

3 (2) Anal segment apparently bipartite in dorsal view, bilobed posteriorly; posterior spiracular opening surrounded by fine float-hairs (Figs. 300, 302) (Clitellariinae p.p.) ...
.... Nemotelus Geoffr. (p. 89)

\- Anal segment not bipartite, at most slightly bilobed posteriorly; posterior spiracular opening without float-hairs........... 4

4 (3) Mature larvae large and stout, about 25-32 mm in length; setae on body segments relatively short and dilated, with frayed margins; anal segment almost transversely oblong (Fig. 303) (Clitellariinae p.p.) .... Clitellaria Meig. (p. 99)

\- Mature larvae much smaller (maximum length about 12 mm); setae less dilated, at most pubescent; anal segment nearly semicircular 5

5 (4) Cf2 setae on head at the level of the eye prominences, and L setae in middle between antennae and eyes (Fig. 42); only 4 pairs of ventral setae on anal segment (Figs. 121, 124, 148) (Sarginae) 6

\- Cf2 setae on head well in front of the eye prominence, and L setae closer to the eyes (Fig. 49); 5 pairs of ventral setae on anal segment (Figs. 397, 400) (Pachygasterinae)........... 8

6 (5) Setae on abdominal segments short, hardly half as long as the segment; setae on anal segment much shorter than the length of penultimate segment (Figs. 148-149) ..
.... Sargus Fabr. (p. 55)

\- Setae on abdominal segments longer, about as long as the segments (Fig. 119); setae on anal segment often longer than penultimate segment (Figs. 121, 124) 7

7 (6) Anal segment with prominent apical lobes; Ap setae as long as Sa setae (Figs. 120-121) .... Chloromyia Duncan (p. 64)

\- Apical lobes on anal segment less distinct; Ap setae strikingly shorter than Sa setae (Figs. 123-125)..... Microchrysa Loew (p. 48)

8 (5) Lateral teeth at anal slit prominent and distinct (Fig. 402)
.... Berkshiria Johnson (p. 128)

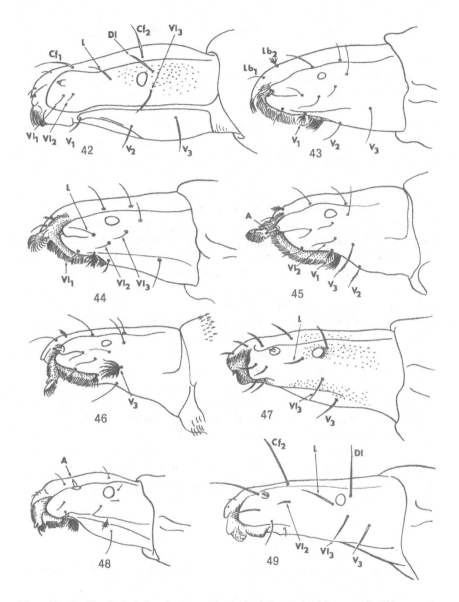

Figs. 42-49. Heads in lateral view of larvae of Stratiomyidae. - 42: Chloromyia formosa (Scop.); 43: Stratiomys chamaeleon (L.); 44: Stratiomys longicornis (Scop.); 45: Odontomyia ornata (Meig.); 46: Oplodontha viridula (Fabr.); 47: Clitellaria ephippium (Fabr.); 48: Oxycera pardalina Meig.; 49: Zabrachia minutissima (Zett.). Abbreviations: see p.132.

-      Lateral teeth at anal slit almost indistinct or absent
       (Figs.397, 400) 9

9 (8) D1 setae on abdominal segments very long, about 1.5-2
       times as long as the segment (Fig.397) 10

-      D1 setae on abdominal segments relatively short, hardly
       as long as the segment (Fig.400) 11

10 (9) Outer dorsal setae (D3) very short, 1/4 as long as the
       segment .... Eupachygaster Kert. (p.125)

-      All dorsal setae about the same length (Fig.396) ....
       .... Zabrachia Coq. (p.127)

11 (9) Ap setae extremely short, virtually invisible in dorsal
       view; Sa setae shorter than L setae of anal segment (Fig.
       400)................................ Neopachygaster Aust. (p.123)

-      Ap setae longer, clearly visible in dorsal view; Sa setae
       mostly longer than L setae of anal segment 12

12 (11) Dorsal setae remarkably thickened or clavate ....
       .... Pachygaster Meig. (p.119)

-      Dorsal setae not thickened or clavate .... Praomyia Kert. (p.121)

13 (2) Antennae situated at anterolateral angles of head capsule
       (Figs. 43-46); accessory anteroventral setae (Av) not
       distinct (Fig.191) (Stratiomyinae) 14

-      Antennae situated between anteroventral angles of head
       capsule and eye prominences (Fig.47); accessory Av seta
       often distinct (Fig.343) (Clitellariinae p.p.)... Oxycera Meig. (p.100)

14 (13) Anal segment elongate, tube-like (Fig.192); posterior
       margins of abdominal segments always without ventral
       hooks (Figs.191, 196) .... Stratiomys Geoffr. (p.67)

-      Anal segment oblong or slightly conical, or rarely re-
       markably elongate then penultimate segment with strong
       ventral hooks (Figs.231, 233) 15

15 (14) V3 seta on head bush-like, with numerous pubescent
       branches (Fig.46); abdominal segments covered dorsally
       with short scale-shaped, apically dilated hairs (Fig.234)
       .... Oplodontha Rond. (p.87)

-      V3 seta on head simple or pubescent (Fig.45); abdominal
       segments without apically dilated hairs (Fig.229).......
       .... Odontomyia Meig. (p.76)

30

Subfamily Beridinae

Antennae elongate, third segment consisting of 8 more or less distinct flagellomeres. Probably with 4 M-veins originally, but M3 partly reduced or absent. Scutellum with 4-8 haired marginal spines. Male genitalia rather simple. Dististyles sometimes with inner lobe, medial process of synsternite either distinct and bilobed or absent. In Europe, only the larvae of Beris spp. are known at present.

Genus Beris Latreille, 1802

Beris Latreille, 1802, Hist.nat.Crust.et Ins., 3: 447.
Type-species: Stratiomys sexdentata Fabricius = Musca chalybeata Forster.

Relatively small species with reduced palpi, conspicuously spined scutellum and elongate abdomen consisting of 7 main segments. Eyes large, meeting on frons and densely haired in males, but well separated and more sparsely and shortly haired in females. Hind metatarsi always thickened in males. Wings often brownish to blackish tinged, with contrasting dark stigma in some species. Only 3 M-veins well-developed. Male genitalia: epandrium usually narrow, with distinct surstyles in the male of fuscipes. Synsternite with more or less prominent posterior medial process. The slender aedeagal complex articulates with symmetrical strip-like dorsal processes of synsternite.

Larvae are characterized by transverse rows of tufts that replace the usual constant setae on body segments. Anal segment rounded posteriorly and fringed with fine setae. Puparia with distinct pupal spiracles on abdominal segments. The description of some larvae published by Lenz (1923) appear to be quite inadequate. Larvae were found under wet decaying bark and in moss cushions near springs, in larval galleries of Cheilosia canicularis Panz. (Syrphidae) in Petasites sp.div., and at the roots of hogweed (Angelica). Adults occur on foliage and ground-vegetation, especially near water.

The genus is represented by 9 species in Europe, 6 of which have been reliably recorded from Denmark and Fennoscandia. The presumed occurrence of B.geniculata Curt. has not yet been confirmed by actual material. Older records of this species in the literature (Wahlberg, 1854; Frey, 1911) were based on misidentifications and in fact refer to fuscipes or strobli (cf. also Andersson,

1971a). The figures of <u>geniculata</u> (Figs. 61-63, 97-100) given here were made from material from Great Britain kindly lent to me by Mr. K. G. V. Smith who bred some specimens from larvae collected at the roots of <u>Angelica</u> near Calne, Wiltshire (Smith, 1957).

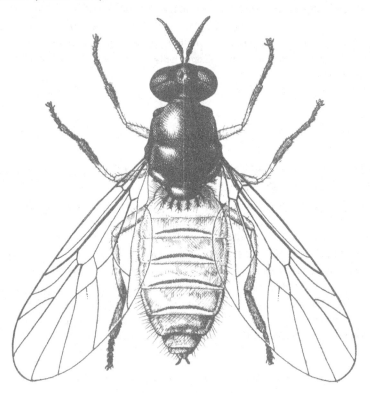

Fig. 50. Female of <u>Beris clavipes</u> (L.).

Key to species of <u>Beris</u>

Adults

1   Ground-colour of abdomen orange (Fig. 50) 2
-       Ground-colour of abdomen brown or black 3
2 (1)   Abdomen with a dark brown preapical transverse stripe
        on each segment (Fig. 50); abdominal pubescence pre-
        dominantly yellow; hind tibiae yellow, hardly darkened
        apically (Fig. 55) .... 2. <u>clavipes</u> (L.) ♂♀

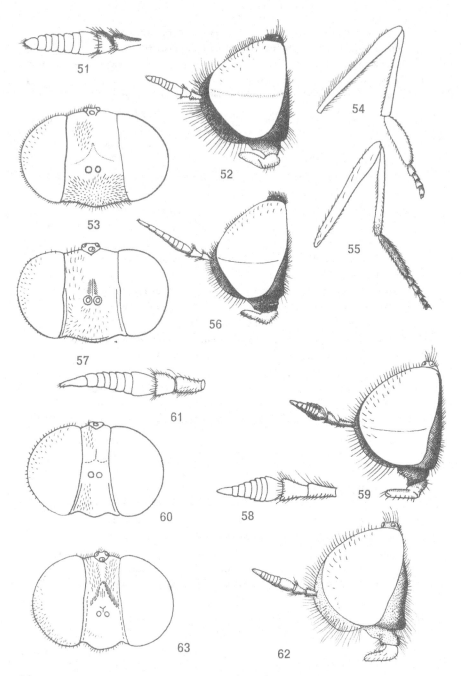

51

52

53

54

55

56

57

58

59

60

61

62

63

34

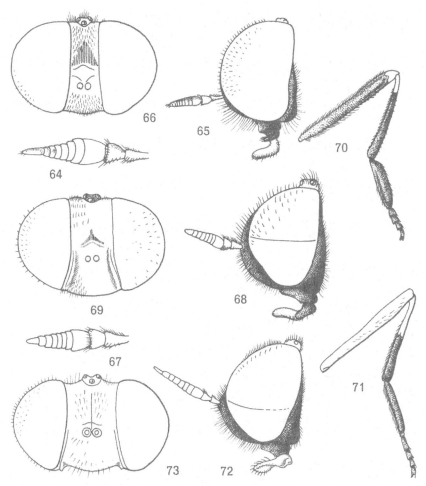

Figs. 64-73. Morphology of adult Beris. - 64: Antenna of B.morrisii Dale; 65: Male head in lateral view of same; 66: Female head in frontal view of same; 67: Antenna of B.strobli Dušek & Rozkošný; 68: Male head in lateral view of same; 69: Female frons in frontal view of same; 70: Hind leg of male of same; 71: Hind leg of female of B.vallata (Forst.); 72: Male head in lateral view of same; 73: Female head in frontal view of same.

Figs. 51-63. Morphology of adult Beris. - 51: Female antenna of B.chalybeata (Forst.); 52: Male head in lateral view of same; 53: Female head in frontal view of same; 54: Hind leg of male of same; 55: Hind leg of female of B.clavipes (L.); 56: Male head in lateral view of same; 57: Female head in frontal view of same; 58: Female antenna of B.fuscipes Meig.; 59: Male head in lateral view of same; 60: Female head in frontal view of same; 61: Antenna of B.geniculata Curt.; 62: Male head in lateral view of same; 63: Female head in frontal view of same.

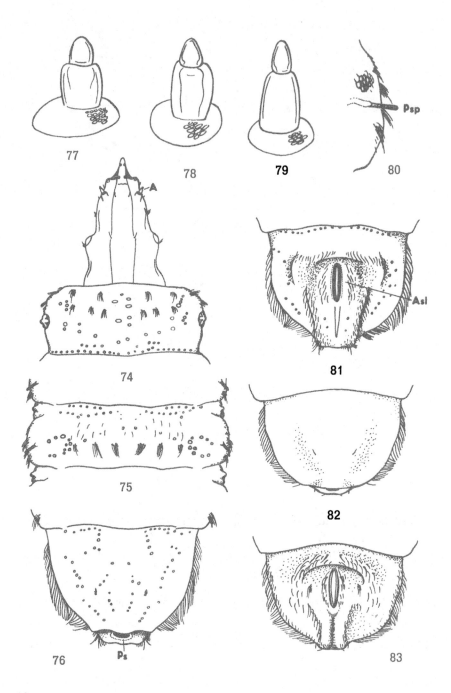

77

78

79

Psp

80

A

74

81

Asl

75

82

76

Ps

83

36

- Femora and tibiae partly brown, but when yellow then
  hind coxae black and frons 1/3 of head-width 10
10 (9) Third antennal segment thickened basally, as long as the
  two basal segments (Fig.58); frons about 1/5 of head-
  width (Fig.60) .... 3. fuscipes Meig. ♀
- Third antennal segment not remarkably thickened in
  basal part; frons wider 11
11 (10) Third antennal segment twice as long as the two basal
  segments (Fig.61); frons about 1/4 of head-width (Fig.63)
  .... geniculata Curt. ♀
- Third antennal segment apparently shorter; frons almost
  1/3 of head-width (Fig.69) .... 5. strobli Dušek & Rozkošný ♀

Larvae and puparia

1 Ventral part of anal segment with striking projections
  posteromedially (Fig.81); antenna relatively short, the
  basal segment at most 1.5 times as long as wide (Fig.77-
  78); pupal spiracles on abdominal segments as long as
  lateral tufts of setae (Fig.80) 2
- Ventral part of anal segment only slightly projecting posteromedially (Fig.83); basal antennal segment twice as
  long as wide (Fig.79); pupal spiracles strikingly shorter
  than lateral tufts of setae .... 4. morrisii Dale
2 (1) Basal antennal segment almost as long as wide (Fig.77);
  marginal setae on anal segment moderately long and
  dense (Fig.81) .... 2. clavipes (L.)
- Basal antennal segment about 1.5 times as long as wide
  (Fig.78); marginal setae on anal segment shorter and
  sparser .... 6. vallata (Forst.)

Figs.74-83. Morphology of Beris-larvae. - 74:Head and thoracic segment 1 in
dorsal view of B.clavipes (L.); 75: Abdominal segment 3 in dorsal view of
same; 76: Anal segment in dorsal view of same; 77: Antenna of same; 78: Antenna of B.vallata (Forst.); 79: Antenna of B.morrisii Dale; 80: Lateral wall
of abdominal segment with pupal spiracle of B.clavipes (L.); 81: Anal segment
in ventral view of same; 82: Anal segment in dorsal view of B.morrisii Dale;
83: Anal segment in ventral view of same. Abbreviations: see p.132.

1. BERIS CHALYBEATA (Forster, 1771)
Figs. 51-54, 84-87.

Musca chalybeata Forster, 1771: 95.
Stratiomys sexdentata Fabricius, 1781: 418.

Dark and rather small species with relatively short antennae, blackish tinge on wings in both sexes and dark abdomen. Frons in female mostly wider than 1/3 of head-width. Pubescence on frons longer than the two basal antennal segments in the male. Legs yellow to dark brown with darkened tarsi, fore and middle tarsi completely dark brown. The intensity of wing darkening is rather variable and can be very weak in females. Colour of abdomen always shining brown in females and dark brown to black in males. - Male genitalia: Medial process of synsternite with a deep median incision posteriorly and two lateral ridges. Aedeagal complex slender and not enlarged apically. - Length: body 5.0-6.0 mm, wing 4.2-5.0 mm.

The larva has not yet been described.

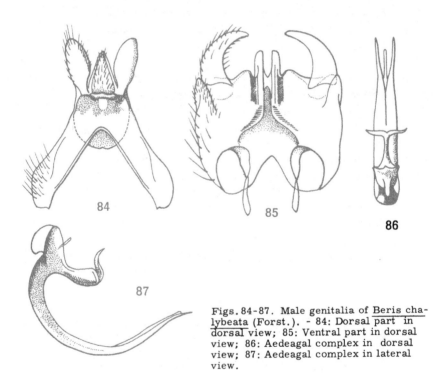

Figs. 84-87. Male genitalia of Beris chalybeata (Forst.). - 84: Dorsal part in dorsal view; 85: Ventral part in dorsal view; 86: Aedeagal complex in dorsal view; 87: Aedeagal complex in lateral view.

The available distribution data include localities from S. Denmark to Finnmark in Norway, Ly. Lpm. in Sweden and from S. Finland to northern Tavastia. - The species is recorded from the greater part of Europe, from France, Italy and Bulgaria to N. Scandinavia. Specimens mentioned by Pleske (1930) from the Far East need to be verified. - May-September. - The larva has been bred from moss according to some older records (cf. Verrall, 1909). Adults are quite common in sheltered valleys and margins of forests and along streams and rivers on low herbage.

Note: Two syntypes of sexdentata Fabricius are preserved in the British Museum (Natural History) (see Zimsen, 1964). Zetterstedt (1842) recorded a male from Västergötland under the name of B. obscura Meigen. This specimen undoubtedly belongs to chalybeata, as has been proved by a recent examination (Andersson, 1971a). B. oscura Meigen is generally considered to be a synonym of chalybeata but, for the time being, the original type has not been studied.

2. BERIS CLAVIPES (Linné, 1767)
    Figs. 35, 50, 55-57, 74-77, 80-81, 88-91.

Musca clavipes Linné, 1767: 981.

Female frons occupying about 1/3 of head-width; eye-margins usually with a slight incision at the level of antennae. Tips of hind tibiae often yellow, as are the whole of the tibiae, but sometimes intensively darkened though never as extensively as in vallata. Wings smoky blackish in both sexes. Abdominal pubescence predominantly yellow, but sometimes also with some erect black hairs. Abdomen normally with dark transverse stripes, but almost completely orange in some males from Finland, and with entirely black first abdominal segment in some specimens from Norway. - Male genitalia: Medial process of synsternite longer than that in vallata, with a deeper median incision. Parameres moderately divergent towards tips. - Length: body 5.5-7.0 mm, wing 4.8-6.0 mm.

The larva was inadequately described by Lenz (1923). Yellowish to dark brown, with some darker cuticular spots forming an inconspicuous pattern. Setae distinctly grouped in transverse rows of tufts on body segments. Posterior spiracular opening on dorsal side of posteromedian projection. Marginal setae on anal segment moderately long and dense. Pupal spiracles on abdominal segments rod-like, as long as tufts of setae on lateral wall. Very similar to larva of vallata, but differing by having lateral setae on abdominal segments and marginal setae on anal segment apparently longer, denser but finer; basal antennal segment shorter; and some differences also in the shape of mandibulomaxillary complex. - Length: 8.0 mm, maximum width: 2.0 mm. (One puparium from Central Europe).

Not a very common species in Denmark and Fennoscandia. Localities are scattered from S.Denmark to NT in Norway, Nb.in Sweden and southernmost Finland. - A European species, well-known from Italy and Bulgaria to C. Scandinavia and the region of Leningrad. - May-July. - Larvae occur in moss and decaying vegetable matter near streams and other water. Adults on leaves of shrubs and low herbage near the larval habitats.

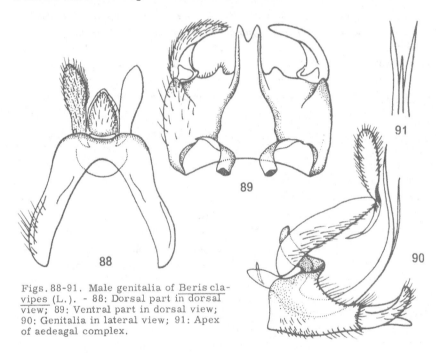

Figs.88-91. Male genitalia of Beris clavipes (L.). - 88: Dorsal part in dorsal view; 89: Ventral part in dorsal view; 90: Genitalia in lateral view; 91: Apex of aedeagal complex.

3. BERIS FUSCIPES Meigen, 1820
Figs. 58-60, 92-96.

Beris fuscipes Meigen, 1820: 8.

A species characterized by the remarkably short and basally thickened third antennal segment, relatively narrow female frons (about 1/5 of the head-width), and unique male genitalia. Number of scutellar spines rather variable (from 4 to 8). Legs bicoloured, brown with yellow tips to femora and bases to tibiae, but often also basal parts of femora and bases of metatarsi yellowish especially in females. Wings only yellowish-brown tinged. - Male genitalia: Epandrium large, with distinct surstyles. Dististyles relatively small, on a solid synsternite, medial process of synsternite reduced, aedeagal complex long and slender. - Length: body 6.2-7.0 mm, wing 5.0-5.8 mm.

The larva was briefly described by Lenz (1923) and seems to be similar to that of clavipes. According to Lenz, the larva of fuscipes should have longer setae on abdominal segments.

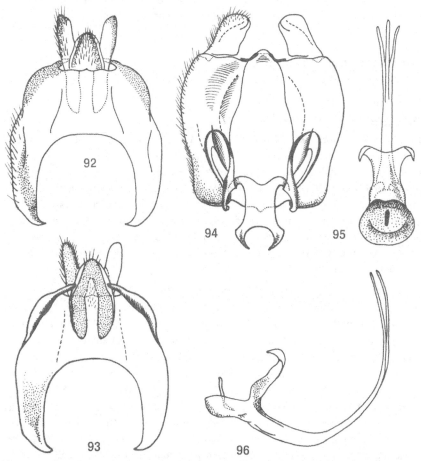

Figs. 92-96. Male genitalia of Beris fuscipes Meig. - 92: Dorsal part in dorsal view; 93: Dorsal part in ventral view; 94: Ventral part in dorsal view; 95: Aedeagal complex in dorsal view; 96: Aedeagal complex in lateral view.

The distribution of the species in Fennoscandia is insufficiently known. Some earlier authors confused this species with geniculata (e.g. Wahlberg, 1854) or mixed it up with strobli (Zetterstedt, 1849; Frey, 1911). B. fuscipes was first recorded from Norway by Andersson (1971a), from Narvik and Vallan. Reliable Swedish localities lie only in Ly. and Lu. Lpm., and in Finland fuscipes has been noted in Kaavi, Kb. Further localities have been recorded from Soviet Carelia (Swir, Suistamo). No records are available from Denmark. - The whole of Europe, and probably also in Siberia. - May-July. - The larva was found under the bark of a tree-trunk lying in a stream. Adults occur in sheltered valleys, often on grass and foliage near running water.

4. BERIS MORRISII Dale, 1842
Figs. 64-66, 79, 82-83, 101-105.

Beris morrisii Dale, 1842: 175.
Beris pallipes Loew, 1846a: 284.

Larger species with pale yellow legs and brownish tarsi. Wings light yellow with brown stigma. Antennae inserted strikingly far below the middle of head-profile, especially in males. Frons of female narrow, rather as in fuscipes (often even less than 1/5 of head-width). Unlike chalybeata fore and middle metatarsi mainly yellow, and female hind coxae usually also yellowish. - Male genitalia: Epandrium almost semicircular, medial process of synsternite prominent and bipartite apically, tips somewhat enlarged. Dististyles large, curved proximally. Aedeagal complex stout and relatively short. - Length: body 6.0-7.6 mm, wing 5.2-6.2 mm.

The larva has been described by Dušek & Rozkošný (1963). Uniformly greyish brown and resembling clavipes as regards external morphology, but anal segment distinctly less prominent posteriorly and its marginal fringe shorter. Pupal spiracles only about half as long as lateral setae. - Length: 7.0-9.0 mm, maximum width: 2.0-2.2 mm. (Several puparia from Central Europe).

Not a very common species but widely distributed in Denmark (E. and S. Jutland, Langeland, NE Zealand, Lolland), Sweden (Sk., Sm., Vg.), S. and C. provinces of Finland and Soviet Carelia. So far not found in Norway. - Throughout most of Europe from Italy to C. Scandinavia and from Great Britain to the Leningrad region. - June-August. - Terrestrial larvae occur in decaying organic material, and the larvae described were found in the larval galleries of Cheilosia canicularis Panz. (Syrphidae) in Petasites sp. div. Adults often on leaves of Petasites and other vegetation along streams and rivers.

42

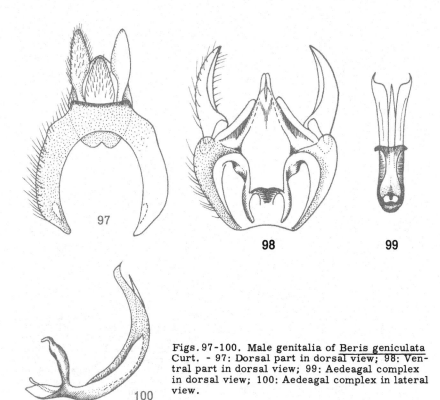

Figs. 97-100. Male genitalia of Beris geniculata Curt. - 97: Dorsal part in dorsal view; 98: Ventral part in dorsal view; 99: Aedeagal complex in dorsal view; 100: Aedeagal complex in lateral view.

5. BERIS STROBLI Dušek & Rozkošný, 1968
Figs. 67-70, 106-109.

Beris chalybeata var. obscura Strobl, 1909: 47.
Beris strobli Dušek & Rozkošný, 1968: 294.

Confused with chalybeata and fuscipes by earlier authors. Size and general appearance similar to chalybeata in particular. Pubescence on frons and upper face hardly as long as the two basal antennal segments in the male. Legs usually bicoloured, brown with broadly yellow knees, but some females from Finland with legs almost entirely yellow. In every case, females differing from chalybeata by predominantly pale yellow fore and middle metatarsi and somewhat narrower frons on upper part, and from morrisii by the obviously wider frons and consistently black hind coxae. - Male genitalia: Medial process of synsternite relatively long, without any lateral ridges. Aedeagal complex with

43

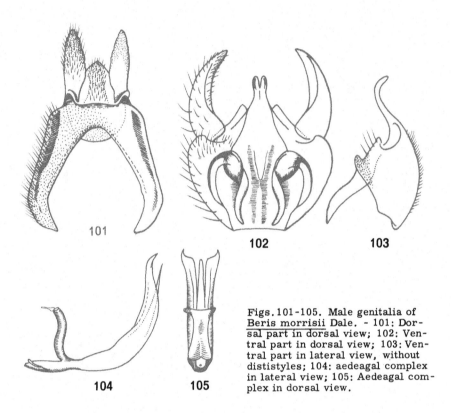

Figs. 101-105. Male genitalia of Beris morrisii Dale. - 101: Dorsal part in dorsal view; 102: Ventral part in dorsal view; 103: Ventral part in lateral view, without dististyles; 104: aedeagal complex in lateral view; 105: Aedeagal complex in dorsal view.

divergent parameres and short aedeagus. - Length: body 4.8-6.5 mm, wing 4.8-5.2 mm.

The larva is not known.

The species is not very rare in Fennoscandia even though it has only been recognized recently (Dušek & Rozkošný, 1968; Andersson, 1971a). The recorded localities in Sweden (from Uppland to Lappmark) and in eastern Fennoscandia (from S. Finland to the Lappish and Carelian provinces in N.) show that it must be widely distributed also in North Europe as well. - C. and N. Europe (Austria, Hungary, Czechoslovakia, Germany, Sweden, eastern Fennoscandia). - June-August. - Adults along streams and the sheltered margins of forests.

6. BERIS VALLATA (Forster, 1771)
    Figs. 71-73, 78, 110-112.

Musca vallata Forster, 1771: 96.

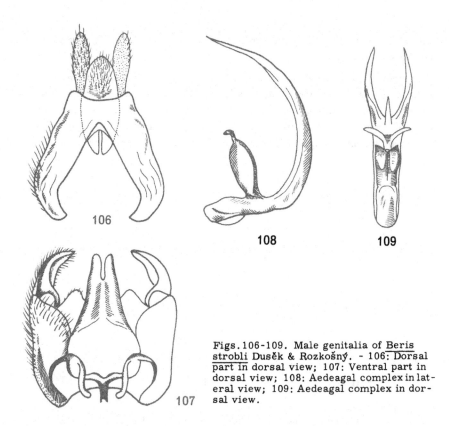

Figs. 106-109. Male genitalia of Beris strobli Dusěk & Rozkošný. - 106: Dorsal part in dorsal view; 107: Ventral part in dorsal view; 108: Aedeagal complex in lateral view; 109: Aedeagal complex in dorsal view.

A species distinguished from the closely related clavipes by having the wings more conspicuously smoky in males only but light yellow tinged in females, the distal halves of hind tibiae black, and no dark preapical transverse stripes on abdominal tergites. Frons of females 1/3 of head-width, and eye-margins always without an incision at level of antennae. Abdomen orange in colour and mainly black haired in males. - Male genitalia: Medial process of synsternite relatively short, with a shallower median incision than in clavipes. Aedeagal apodeme distinct. - Length: body 5.4-6.0 mm, wing 5.0-5.5 mm.

The larva was briefly described by de Meijere (1916), Lenz (1923), Bischoff (1925) and McFadden (1967). It is very similar to that of clavipes, but some differences mentioned by Lenz (1923) are not reliable. The shape of the antennae seems to be diagnostic, and the lateral setae on abdominal segments are stronger and sparser than in clavipes and marginal setae on anal segment are somewhat shorter. - Length: 6.2 mm, maximum width: 1.5 mm. (One puparium from Central Europe).

A rather common species in almost all provinces of Denmark except N.part of Jutland; rare in Sweden (some localities in Sk. and Öl.); not known from Norway; and apparenly quite absent in eastern Fennoscandia. - Most of Europe but mainly in C. and N. parts. - June-August. - The larvae live in decaying leaves and wet moss, and the adults on vegetation near water, leaves of bushes, etc.

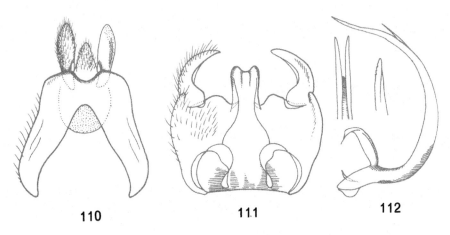

110 **111** **112**

Figs.110-112. Male genitalia of Beris vallata (Forst.). - 110: Dorsal part in dorsal view; 111: Ventral part in dorsal view; 112: Aedeagal complex in lateral view, its apex and tip of paramere.

Genus Exodontha Rondani, 1856

Exodontha Rondani, 1856, Dipt.Ital.Prodr., 1:169.
Hexodonta, emend.
Acanthomyia Schiner, 1860, Wien.ent.Monatschr., 4: 2.
Type-species: Exodontha pedemontana Rondani, 1856 = Beris dubia Zetterstedt, 1838.

Robust species with well-developed, two-segmented palpi and broad abdomen consisting of five main segments only. Eyes densely haired in both sexes. One apical spur on middle tibia. Wing long and broad, discal cell large, M3 partly reduced but present, M4 arising from discal cell.

46

Three species in the World, with two North American species in addition to the Palearctic dubia. The larva of a North American species is about 13-15 mm long and was found in moist rotting wood under large boulders on a mountain-side (McFadden, 1967). This habitat corresponds rather well with the known localities of dubia in Europe.

7. EXODONTHA DUBIA (Zetterstedt, 1838)
Figs. 34, 113-117.

Beris dubia Zetterstedt, 1838: 512.

The largest species of the European Beridinae. Some features appear to be variable, e.g. number of scutellar spines (4 to 8), or wing-venation where bases of M1 and M2 may be touching, separated or united into a common stem. In addition to the characters mentioned above, the male genitalia are very specific: Epandrium rather large, almost semicircular, proctiger stout, with oval cerci. Synsternite without medial process but with distinct dorsal bridge, dististyles bilobed. Aedeagal complex short and broad, parameres strong and finely spinulose on lateral and ventral sides. - Length: body 7.5-10.5 mm, wing 6.2-7.0 mm.

The larva is not known.

The species was described from Dovre (Norway) by Zetterstedt (1838) and was subsequently also captured in Ås near Östersund, Jmt., Tärna, Ly.Lpm. and Kvikkjokk, Lu.Lpm. in Sweden. Material has also been seen from Bergen, HOi and Amodt, HEn in Norway. - Mainly mountainous regions in C.Europe, Scandinavia, the Altai, and recently found in Japan (Nagatomi, 1968). - July-August. - In shady and mainly montane woods, up to 1200 m in Swedish Lapland.

Subfamily Sarginae

Antennae short or moderately long, third segment consisting only of 4 main flagellomeres and a subterminal arista. Male eyes usually also dichoptic, except for Microchrysa, densely haired in Chloromyia. Scutellum always unarmed. Leg colour often enormously variable, from entirely yellow to almost black in some species. 4 M-veins always developed, and a distinct cross-vein present between M4 and discal cell. Male genitalia rather simple, but specific. Epandrium with surstyles in Microchrysa; a low medial process on synsternite usually more or less developed, but quite absent in Microchrysa.

47

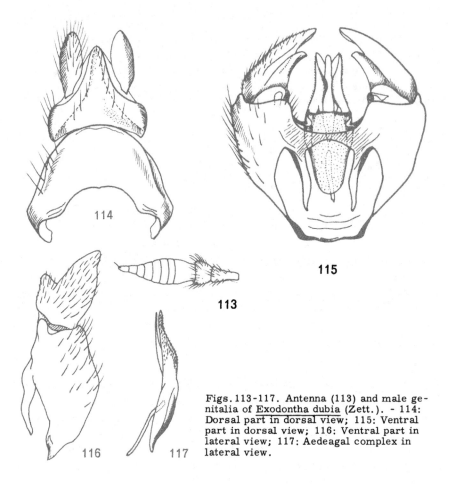

Figs. 113-117. Antenna (113) and male genitalia of Exodontha dubia (Zett.). - 114: Dorsal part in dorsal view; 115: Ventral part in dorsal view; 116: Ventral part in lateral view; 117: Aedeagal complex in lateral view.

The larvae of Sarginae have been dealt with by Brindle (1965) in considerable detail.

Genus Microchrysa Loew, 1855

Microchrysa Loew, 1855, Verh. zool. -bot. Ges. Wien, 5: 146.
Type-species: Musca polita Linné, 1758.

Tiny, shining green, blue or black species with bare or indistinctly haired eyes which are touching in males and widely separated in females. Anal cell on wings rather wide. Male genitalia bearing short, ventrally and inwardly directed surstyles, and a posteriorly notched synsternite without medial process. Dististyles with low inner lobe in some species. Aedeagal complex stout in distal part, with duplicated parameres and apically tripartite aedeagus.

Larvae are the smallest in the Sarginae (about 7 mm long) but have relatively long bristles. In particular, Lb and Cf setae on the head are conspicuously long. Anal segment slightly bilobed apically, and apical setae (Ap) usually very short. Puparia without distinct pupal spiracles on abdominal segments. Larvae breed in dung, decaying vegetable matter and compost.

The genus is distributed in all the main zoogeographical regions. 2 of the 3 European species were introduced into North America.

Key to species of Microchrysa

Adults

1   At least basal antennal segments yellowish 2
-    Antennae completely black; notopleural line darkened and
     very narrow .... 10. polita (L.)
2 (1) Abdomen in both sexes and female frons shining black; noto-
     pleural line brownish, not broadened before wing-base ...
     .... 8. cyaneiventris (Zett.)
-    Abdomen and female frons metallic green or blue; noto-
     pleural line strikingly yellow, broadened before wing-base
     in particular .... 9. flavicornis (Meig.)

Larvae

1   Dorsal and ventral setae on body segments with expanded
     tips .... 8. cyaneiventris (Zett.)
-    Dorsal and ventral setae without remarkably expanded tips 2
2 (1) Median ventral setae (V1) on abdominal segment 6 (at ster-
     nal patch) distinctly smaller then other ventral setae; Ap
     setae on anal segment only 1/4 as long as Sa setae (Fig.
     125) .... 10. polita (L.)
-    Median ventral setae on abdominal segment 6 as long as
     other ventral setae; Ap setae about 1/3 as long as Sa setae
     (Fig.123) .... 9. flavicornis (Meig.)

49

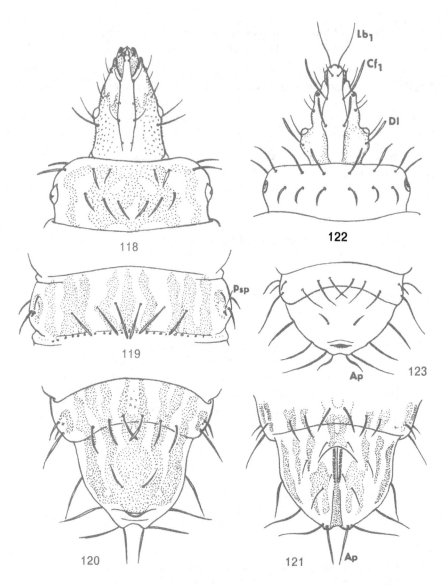

Figs. 118-123. Morphology of larvae of Sarginae. - 118: Head and thoracic segment 1 in dorsal view of <u>Chloromyia formosa</u> (Scop.); 119: Abdominal segment 3 in dorsal view of same; 120: Anal segment in dorsal view of same; 121: Anal segment in ventral view of same; 122: Head and thoracic segment 1 in dorsal view of <u>Microchrysa polita</u> (L.); 123: Posterior end in dorsal view of <u>Microchrysa flavicornis</u> (Meig.). Abbreviations: see p. 132.

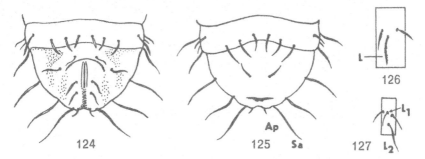

Figs. 124-127. Morphology of larvae of Sarginae. - 124: Posterior end in ventral view of Microchrysa polita (L.); 125: Posterior end in dorsal view of same; 126: Scheme of setae on lateral wall of abdominal segment 3 of Chloromyia formosa (Scop.); 127: Same of Microchrysa polita (L.). Abbreviations: see p.132.

8. MICROCHRYSA CYANEIVENTRIS (Zetterstedt, 1842)
Figs. 128-132.

Chrysomyia cyaneiventris Zetterstedt, 1842: 156.

Antennae always yellow, at most only the last flagellomere of third antennal segment darkened. Narrow male frons black and densely covered with adpressed grey hairs. Female frons shining black. Fore legs mainly pale yellow, black pattern on middle and hind legs also reduced and often only femora darkened in middle. Notopleural line brownish, less distinct than in flavicornis and not broadened in front of wing-base. Abdomen shining black, at most with a bronze tinge, sometimes brownish translucent at margin. - Male genitalia: The shape of aedeagal complex is particularly characteristic. Parameres with a simple point apically. Dististyles large and flat, almost leaf-shaped. - Length: body 3.0-4.3 mm, wing 3.5-4.0 mm.

The larva has been briefly described by Brindle (1965). It is smaller than that of polita and lighter in colour due to the wider lighter longitudinal bands. Tips of dorsal and ventral setae distinctly expanded. Length about 6 mm.

Described by Zetterstedt (1842) from several Swedish localities, and recently discovered in Bl., Sm., Gtl., and Dlr. Also found in Denmark (E. and S. Jutland, NE Zealand), Norway (1 male from Bland Skoy, HOy) and Finland (several provinces, from S. Finland to Ostrobottnia). It is probably not very rare in northern Europe. - W. and N. Europe, and rarely also in Czechoslovakia, but older records need verification. - June-July, rarely also September. The larva has been found in the soil beneath moss on tree trunks, but possibly has a fairly wide range of habitats like polita.

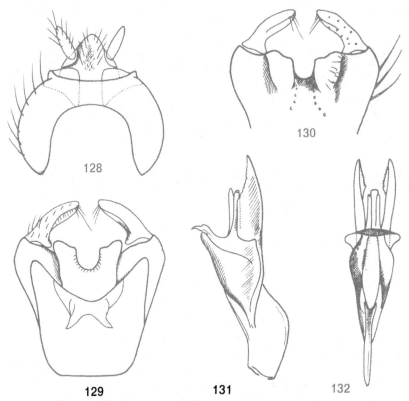

Figs. 128-132. Male genitalia of Microchrysa cyaneiventris (Zett.). - 128:Dorsal part in dorsal view; 129: Ventral part in dorsal view; 130: Ventral part in ventral view; 131: Aedeagal complex in lateral view; 132: Aedeagal complex in dorsal view.

9. MICROCHRYSA FLAVICORNIS (Meigen, 1822)

Figs. 123, 133-136.

Sargus flavicornis Meigen, 1822: 112.

Sargus pallipes Meigen, 1830: 344.

Chrysomyia pallipes Meigen; Zetterstedt, 1842: 156.

Basal antennal segments yellowish, third segment brown. Body light green or blue metallic shining. Notopleural line strikingly yellow, stripe-like but expanded in front of wing-base. Femora usually darkened as well as tips of hind tibiae. Third antennal segment largely yellow and fore legs completely light in some males. - Male genitalia: Synsternite with a deep, rounded notch in middle. Aedeagal complex strongly tapering in basal half, the duplicated apical parts slender and shifted towards the posteroventral angles. - Length: body 3.5-4.8 mm, wing 4.0-4.5 mm.

The larva is yellowish-brown with 3 indistinct darker longitudinal stripes. The main chaetotaxy is as in polita, but usually one pair of ventral bristles on first thoracic segment distinct, all ventral setae on abdominal segment 6 about the same length, and Ap setae 1/3 as long as Sa setae on anal segment. - Length: 5.5 mm, maximum width: 1.5 mm (2 puparia in ZMC).

Rather common in Denmark, S. and C. Fennoscandia though not as frequent as polita. Known from all provinces in Denmark except NWZ. Scattered localities in Fennoscandia, reaching as far north as NT in Norway, to Jmt. in Sweden, and to Finnish Lapland in eastern Fennoscandia. - Very probably throughout Eurosiberia, from Europe through eastern Siberia to the Pacific Ocean. Also introduced into North America. - May-July. - Larvae in decaying organic matter, and probably also in dung; adult coprophily is mentioned by Hammer (1941), Papp (1971) and McFadden (1972). Adults in pastures and gardens, but also in mountains and on the sea-shore.

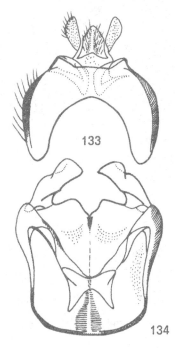

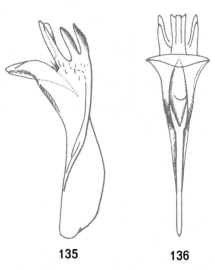

Figs. 133-136. Male genitalia of Microchrysa flavicornis (Meig.). - 133: Dorsal part in dorsal view; 134: Ventral part in dorsal view; 135: Aedeagal complex in lateral view; 136: Aedeagal complex in dorsal view.

10. **MICROCHRYSA POLITA** (Linné, 1758)
    Figs. 122, 124-125, 127, 137-142.

Musca polita Linné, 1758: 598.
Sargus cyaneus Fabricius, 1805: 258.

Antennae entirely black, legs dark brown to black with largely yellow knees, tips of tibiae and bases to metatarsi. Notopleural line hardly distinct, usually brown. The shade of metallic shine and the leg colour are particularly variable. - Male genitalia: Medial notch on posterior margin of synsternite usually as deep as in flavicornis. Aedeagal complex less tapering in basal part, duplicated tips of parameres relatively short and thickened basally, not strikingly shifted towards the posteroventral angles. - Length: body 4.5-5.5 mm, wing 4.8-5.3 mm.

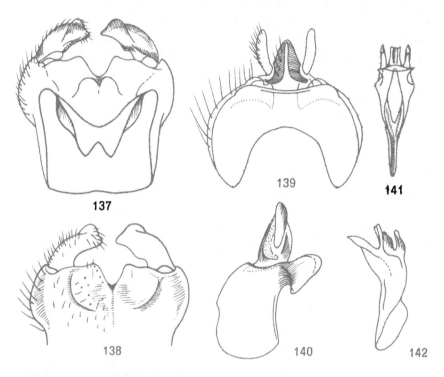

Figs. 137-142. Male genitalia of Microchrysa polita (L.). - 137: Ventral part in dorsal view; 138: Ventral part in ventral view; 139: Dorsal part in dorsal view; 140: Dorsal part in lateral view; 141: Aedeagal complex in dorsal view; 142: Aedeagal complex in lateral view.

The larva has been described by several previous authors (e.g. Bouché, 1834; Müller, 1925; Lobanov, 1960; Brindle, 1965). Yellowish brown, with 3 weakly distinct darker longitudinal stripes dorsally and similar well-developed stripes ventrally. Setae on head mostly long, especially Lb1, Cf and Dl. Tips of dorsal and ventral setae not remarkably expanded. Ap setae of last abdominal segment very short, hardly 1/4 as long as Sa setae. - Length: 7.0-8.0 mm, maximum width: 2.0 mm. (7 puparial exuviae in ZMC).

The commonest species of the genus, even in North Europe. Known to occur in virtually all provinces of Denmark (except WJ), from S. provinces in Norway to Nn, through Sweden and Finland to Lappish provinces and to Soviet Carelia. - Holarctic species distributed all over Europe, W. and C. Asia including Mongolia, and North America. - May-August. - Larvae live in dung, grass heaps, soil beneath moss, etc. Winter diapause in the mature larval stage.

Genus Sargus Fabricius, 1798

Sargus Fabricius, 1798, Ent. Syst., Suppl., 566.
Geosargus Bezzi, 1907, Wien. ent. Ztg, 26: 51.
Type-species: Musca cupraria Linné, 1758.

The European species of the genus are mostly shining metallic green or violet, the male having abdomen with bronze to coppery reflections. Antennae with rather short basal segments, and annulate circular third segment with a subterminal arista. Eyes bare, separated in both sexes but frons narrower in males. Anal cell of wing relatively narrow. Abdomen mainly slender, somewhat narrowed basally. Male epandrium with short surstyles in rufipes and splendens. A higher or lower medial process posteriorly on synsternite as well as two more or less prominent lateral lobes. Aedeagal complex with two symmetrical parameres, aedeagus tripartite, aedeagal apodeme rather strong.

Larvae with relatively short setae on body segments. No apparent differences were given in the published descriptions of the larvae of cuprarius and iridatus by Johannsen (1922), Müller (1925), Lundbeck (1907) and Séguy (1926). Brindle (1965) assumed some diagnostic characters in the shape of head in both species. The larvae of rufipes and splendens are not known at present.

Representatives of the genus are known from all the main zoogeographical regions. 6 species are known to occur in Europe, 4 of which have been found in North Europe. S. rufipes, recently redefined by Andersson (1971b), seems to be a typically boreal species.

Key to species of Sargus

Adults

1   Posterior margin of head fringed with erect pale hairs
    (Fig. 143); epandrium almost transverse-oblong, without
    surstyles (Figs. 152, 155) 2
-    Posterior fringe on head absent (Figs. 144-146); epandrium
    almost elongate-oval, with short surstyles posteroventral-
    ly (Figs. 163, 169) 3
2 (1) Wings with conspicuous dark cloud about middle, below the
    stigma (Fig. 159); aedeagus distinctly tickened (Figs. 153-
    154) .... 11. cuprarius (L.)
-    Wings uniformly smoky without a contrasting middle cloud
    (Fig. 160); aedeagus slender (Figs. 157-158)....... 12. iridatus (Scop.)

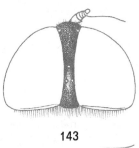

143

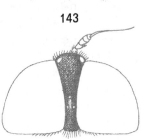

144

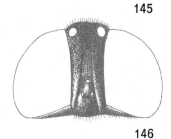

145

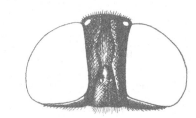

147

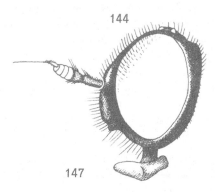

146

Figs. 143-147. Heads of Sarginae. - 143:
Male in dorsal view of Sargus cuprarius
(L.); 144: Female in dorsal view of S.
rufipes Wahlbg.; 145: Male in dorsal
view of S. splendens Meig.; 146: Female
of same; 147: Female in lateral view of
Chloromyia formosa (Scop.).

3 (1) Synsternite of male without accessory posteroventral lobe
(Fig. 162); parameres distinctly longer than aedeagus (Fig.
164); female frons relatively wide, with a broadened bare
midfrontal stripe (Fig. 144) .... 13. _rufipes_ Wahlbg.

\- Synsternite of male with a pair of accessory posteroventral
lobes (Figs. 166-167); parameres as long as aedeagus (Figs.
170-171); female frons much narrower, with the bare mid-
frontal stripe hardly broadened (Fig. 146) .... 14. _splendens_ Meig.

Larvae

1 Head longer than wide at posterior part .... 12. _iridatus_ (Scop.)

\- Head wider than long .... 11. _cuprarius_ (L.)

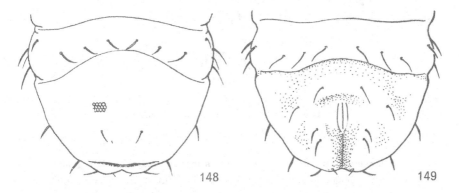

Figs. 148-149. Posterior end of larva of _Sargus iridatus_ (Scop.) in dorsal view
(148) and ventral view (149).

11. SARGUS CUPRARIUS (Linné, 1758)
Figs. 143, 151-154, 159.

Musca cupraria Linné, 1758: 598.
Sargus nubeculosus Zetterstedt, 1842: 157.
Sargus minimus Zetterstedt, 1849: 2965.

A species characterized by a post-ocular fringe and a dark median mark on
wings. Legs usually dark brown to black, with broadly yellow knees and hind
metatarsi either entirely dark or yellow at base. The species is very variable
in size, and remarkably small specimens occur rather often, particularly in N.
Europe. Difference in leg-colour, especially the hind metatarsi, were considered to be of specific value (_nubeculosus_, with completely black hind metatar-

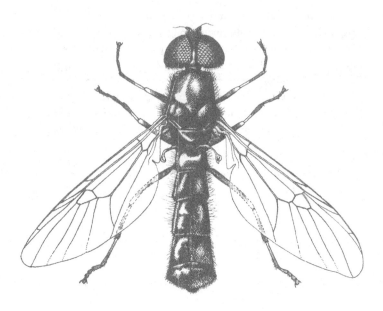

Fig. 150. Male of <u>Sargus splendens</u> Meig.

si). Wing darkening is sometimes hardly distinct in males. - Male genitalia: The thick-set aedeagus distinguishes this species quite reliably from the close-ly-related <u>iridatus</u>. Length: body 6.0-10.5 mm, wing 5.5-9.0 mm.

The larva is very similar to that of <u>iridatus</u>. Brown, with extreme lateral and posterior margin lighter. Setae conspicuously short, hardly longer than half length of abdominal segments. Head relatively short and wide, apparently wider than long according to Müller (1925). - Length: 9.0-12.0 mm, maximum width 2.8 mm.(One puparium without anterior part in ZMO).

Common throughout the greater part of Denmark and Fennoscandia, N to Lappish provinces in Sweden and Finland; apparently less common in Norway (Ø, Ak, STi, TRy).- Holarctic species: Europe, Asia to Mongolia, and North America from S.Canada to California. - May-August. - Larvae in cow dung, compost, garden refuse, etc., and adults frequent, especially in gardens visit-ing various flowers (Daucaceae, Asteraceae, etc.).

Note: The synonymy of <u>nubeculosus</u> and <u>minimus</u> has been verified by Anders-son (1971b) who confirmed it with lectotype designations.

58

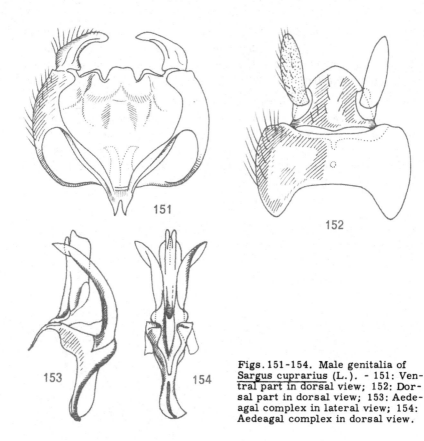

Figs. 151-154. Male genitalia of
Sargus cuprarius (L.). - 151: Ventral part in dorsal view; 152: Dorsal part in dorsal view; 153: Aedeagal complex in lateral view; 154:
Aedeagal complex in dorsal view.

12. SARGUS IRIDATUS (Scopoli, 1763)
    Figs. 33, 36, 148-149, 155-158, 160.

Musca iridata Scopoli, 1763: 340.
Sargus infuscatus Meigen, 1822: 107.

A species with a post-ocular fringe, and uniformly greyish-brown tinged wings
without a conspicuous median cloud. Legs usually entirely dark brown but their
colour is variable as in cuprarius. The intensity of the wing-darkening may also be rather different. - Male genitalia: Compared with cuprarius, synsternite
more rounded, its medial process narrower, and aedeagus more slender. -
Length: body 6.0-11.8 mm, wing 5.5-10.0 mm.

The larva was briefly described by Lundbeck (1907). Greyish-brown, with
more or less distinct darker longitudinal stripes along the dorsum and venter.

Setae on body segments short, 1/3-1/2 length of abdominal segments. Head longer than wide. Ventral setae divergent, and all 4 setae on lateral walls of abdominal segments almost equal. - Length: 10.0-12.0 mm, maximum width 3.0 mm. (One puparium in ZMC).

Very common; appears to be generally more frequent than cuprarius in N. Europe. Widespread in all provinces of Denmark and most provinces of Fennoscandia, reaching as far north as Nn in Norway and the Lappish provinces of Sweden and Finland, also to Soviet Lapland. - Known throughout the whole of Europe. - May-August. - Larvae in decaying grass-heaps, compost, manure, cow dung, etc. Adults in gardens, meadows and pastures.

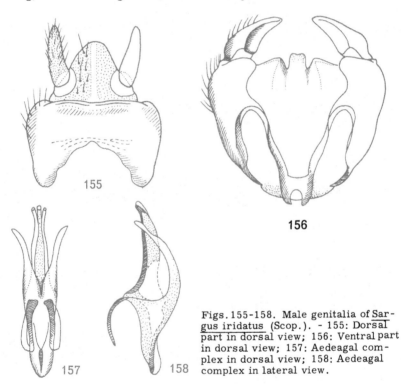

Figs. 155-158. Male genitalia of Sargus iridatus (Scop.). - 155: Dorsal part in dorsal view; 156: Ventral part in dorsal view; 157: Aedeagal complex in dorsal view; 158: Aedeagal complex in lateral view.

13. SARGUS RUFIPES Wahlberg, 1854
    Figs. 144, 161-165.

Sargus rufipes Wahlberg, 1854: 213.

Post-ocular fringe absent. Female head relatively shorter than that of splend-

159

160

Figs. 159-160. Wings of 159: Female of Sargus cuprarius (L.) and 160: Female of S. iridatus (Scop.). x 8.

ens. Bare mid-frontal stripe distinctly broader. White frontal calli small, sometimes not distinctly white. Femora and tibiae completely reddish-yellow, and tarsi darkened. Thoracic hairs black, and abdominal hairs wholly yellowish or black along a more or less broad dorsocentral stripe. - Male genitalia: Posterior margin of synsternite with low medial process and without accessory posteroventral lobes. Aedeagus strikingly shorter than parameres. - Length: body 9.5-11.0, wing 7.8-9.2 mm.

The larva is not known.

A rare boreal species, with reliable records only from Fennoscandia at present. Apart from the type locality (Kvikkjokk, Lu. Lpm., Sweden), the species has been found in some other C. and N. provinces of Sweden (Dlr., Jmt., Ly. Lpm. and T. Lpm.), in the region of Laurgård, On and Harstad, TRy in Norway, in Finnish Lapland (Kilpisjärvi, Le) and Kuusamo (Salla, Kuusamo, Ks), and in Soviet Carelia (Paanajärvi, Kr). - June-July. - Records in the literature from other parts of Europe need verification.

Note: S. rufipes has been regarded as a synonym of splendens by most recent authors (e.g. Dušek & Rozkošný, 1964; Nartshuk, 1969). The full validity of this species has been shown by Andersson (1971b), who designated a lectotype and differentiated it from the closely-related splendens.

14. SARGUS SPLENDENS Meigen, 1804
   Figs. 145-146, 150, 166-171.

Sargus splendens Meigen, 1804: 144.
Sargus flavipes Meigen, 1822: 108.
Sargus nigripes Zetterstedt, 1842: 159.

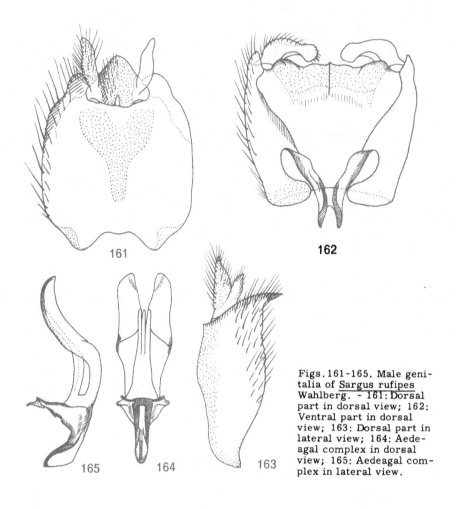

Figs. 161-165. Male genitalia of Sargus rufipes Wahlberg. - 161: Dorsal part in dorsal view; 162: Ventral part in dorsal view; 163: Dorsal part in lateral view; 164: Aedeagal complex in dorsal view; 165: Aedeagal complex in lateral view.

Post-ocular fringe absent. Female head more semicircular in dorsal view, and bare mid-frontal stripe apparently more slender than in rufipes. White lateral frontal calli larger and contrastingly white. Thoracic hairs white or yellow or blackish on central stripe of mesonotum. Leg colour very variable: in males, femora and tibiae seldom entirely reddish-yellow, femora usually darkened in middle, or legs extensively black with yellow knees and bases of metatarsi; in females, legs usually mainly yellow with darkened fore femora and tarsi. Abdominal hairs whitish and long laterally, black and short dorsomedially. - Male genitalia: Posterior margin of synsternite with low medial process, later-

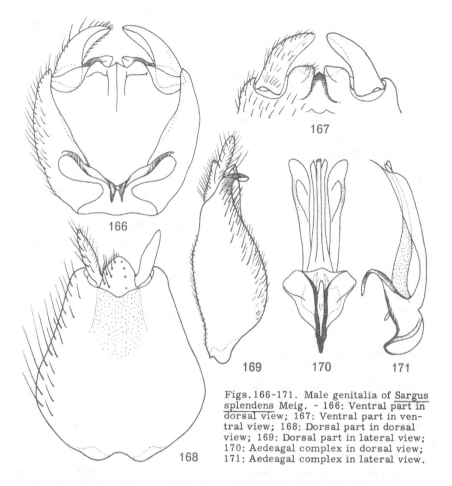

Figs.166-171. Male genitalia of Sargus splendens Meig. - 166: Ventral part in dorsal view; 167: Ventral part in ventral view; 168: Dorsal part in dorsal view; 169: Dorsal part in lateral view; 170: Aedeagal complex in dorsal view; 171: Aedeagal complex in lateral view.

al lobes and a pair of accessory posteroventral lobes. Aedeagus almost as long as parameres. - Length: body 6.5-10.0 mm, wings 6.0-7.8 mm.

The larva is brownish-white according to Lundbeck (1907). Distinguishing characters from the other species of the genus are not known. Length: 8.0-9.0 mm.

Rather widespread in Denmark (from SJ, LFM and B to NWJ and NEZ) as well as in Fennoscandia - from Ø to ST in Norway, in many provinces of Sweden (N to Lapland) and in S. and C. Finland. - C. and N. Europe, Siberia. Apparently more frequent in N. - June-September. - The larva was found in earth. Adults occur on vegetation, on cow dung in pastures, etc.

Note: Andersson (1971b) confirmed the synonymy of nigripes by examining the genitalia of the male holotype.

Genus Chloromyia Duncan, 1837

Chloromyia Duncan, 1837, Mag. Zool. Bot., 1: 145.
Chrysomyia Macquart, 1834, Hist. nat. Ins. Dipt., 1: 262.
Type-species: Musca formosa Scopoli, 1763.

Metallic shining species, with abdomen golden in males and green to violet in females. Third antennal segment suboval with subterminal arista. Eyes with dense and long hairs, touching in males and broadly separated in females. Male aedeagal complex with duplicated and slender parameres.

Larvae with inconspicuous labral setae. Setae on body segments rather long and mostly pubescent. Anal segment bilobed apically, Ap and Sa setae equal. Larvae saprophagous, living in earth and under stones, and adults often on flowers. Two European species, Ch. speciosa (Macq.) occurring only in S. and C. Europe.

15. CHLOROMYIA FORMOSA (Scopoli, 1763)
    Figs. 42, 118-121, 126, 147, 172-177.

Musca formosa Scopoli, 1763: 339.
Musca aurata Fabricius, 1787: 347.

Medium-size species, with remarkable metallic sheen. Head pubescence yellowish-brown, wings light-brownish tinged, and hind tarsi including metatarsi entirely dark brown, at most yellowish pubescent. - Male genitalia: Posterior margin of synsternite with two oval lobes between simple dististyles, and a small median incision. Aedeagal complex with very slender and divergent duplicated parameres. A low aedeagal apodeme present. - Length: body 7.3-9.0 mm, wing 6.0-6.8 mm.

The larva was inadequately described by Cornelius (1860), Lundbeck (1907) and Séguy (1926). Pale yellow, with brownish pattern in the form of 7 more or less distinct longitudinal stripes and dorsolateral spots on each segment. Labral setae short, setae on body segments mostly as long as each segment, pubescent. Only 3 setae on lateral wall of abdominal segments. V1 setae divergent, but other ventral setae convergent. Ap setae of anal segment nearly as long as Sa setae. Puparium usually paler than larva, and dark pattern usually indistinct. Very short, cylindrical pupal spiracles distinct on abdominal segments 2-5 dorso-laterally. - Length: 8.2-12.0 mm, maximum width: 2.0-3.0 mm.

A common species all over Danmark (except WJ) and also frequent in Scandinavia though, for the time being, known only from a few provinces of Norway (HE, Ry, HOy), but widespread in S. and C.Sweden. In Finland only known to occur on Aland Islands. - Widely distributed throughout the Palearctic Region from North Africa to C.Scandinavia and from Asia Minor to Siberia and China. - April-August. - Larvae in earth, under stones, and sometimes also in dung. Overwintering larvae pupate in spring. Adults on low herbage, leaves of shrubs and flowers, in sunny places.

Note: In Fabricius' collection (ZMC) are fragments of 3 specimens (probably 2 males and 1 female), labelled "aurata" in Fabricius' handwriting. All belong without any doubt to formosa.

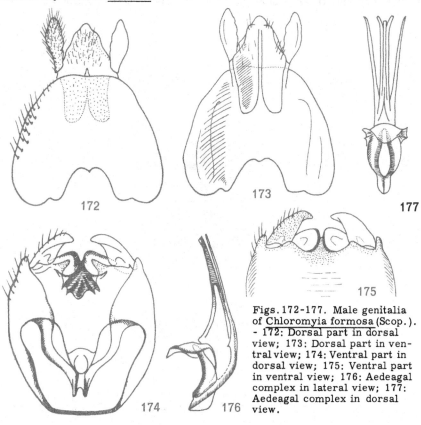

Figs.172-177. Male genitalia of Chloromyia formosa (Scop.). - 172: Dorsal part in dorsal view; 173: Dorsal part in ventral view; 174: Ventral part in dorsal view; 175: Ventral part in ventral view; 176: Aedeagal complex in lateral view; 177: Aedeagal complex in dorsal view.

Subfamily Stratiomyinae

Antennae elongate, third segment consisting of 5-6 flagellomeres without any arista. Scutellum with 2 spines. 4 M-veins originally developed. M4 connected to discal cell by means of a distinct cross-vein. Abdomen consisting of 5 main segments, usually broad and flattened. The basic shape of the male genitalia is essentially uniform, but the form of the dististyles, the medial process of synsternite and the aedeagal complex are specific.

The larvae are mostly aquatic, and are characterized by an apical coronet of float-hairs at apex of anal segment. Antennae always anterior in position. Some larvae bear strong ventral hooks on penultimate or even ante-penultimate segment. The main larval characters of the subfamily were summarized and completed by Brindle (1964a). In North Europe only 11 species in 3 genera.

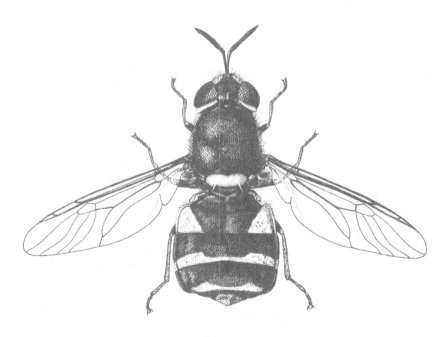

Fig.178. Female of _Stratiomys potamida_ Meig.

Genus Stratiomys Geoffroy, 1762

Stratiomys Geoffroy, 1762, Hist. abrég. Ins., 2: 449, 475.
Type-species: Musca chamaeleon Linné, 1758.

Large species with long antennae, and usually with yellow markings on head (females), scutellum and abdomen. First antennal segment rod-like, at least 4 times as long as second, third segment consisting of 5-6 flagellomeres. Male aedeagal complex with a relatively long, compact basal part.

Larvae elongated, with a conspicuously lengthened, tube-like anal segment which may be 8-10 times longer than wide. No ventral hooks on abdominal segments. On head V1 setae always and Lb2 usually plumate, branched. D2 far in front of other dorsal setae on body segments, and ventral setae arranged in a semicircular row. 6 ventral setae on anal segment. Larvae live in shallow standing water, feeding upon fine detritus (Schremmer, 1951). Only 4 species in North Europe.

Note: Siebke (1863) described the male of a Stratiomys from Norway as S. paludosa new species. This species was later identified as S. validicornis

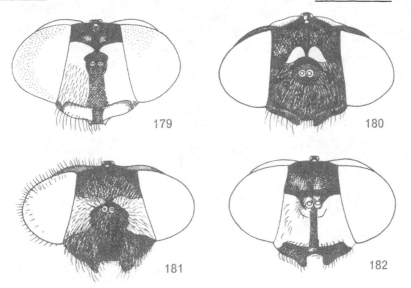

Figs. 179-182. Female heads in frontal view of Stratiomys. - 179: S. chamaeleon (L.); 180: S. furcata Fabr.; 181: S. longicornis (Scop.); 182: S. potamida Meig.

(Loew, 1854). Unfortunately, the original type-material of paludosa is very probably lost. According to present knowledge of the distribution of validicornis, its occurrence in Norway is scarcely likely; at any rate, no actual material is known from Scandinavia at present. S. paludosa might therefore be based on a rather aberrant specimen of furcata.

Key to species of Stratiomys

Adults

1   Dorsum of abdomen black, at most with fine whitish hairs
       at posterior angles of tergites .... 18. longicornis (Scop.)
-      Dorsum of abdomen with yellow markings (Figs. 183-188) 2
2 (1) Yellow lateral markings fused medially on tergite 4 (male)
       or on tergites 3 and 4 (female) (Figs. 187-188)...... 19. potamida Meig.

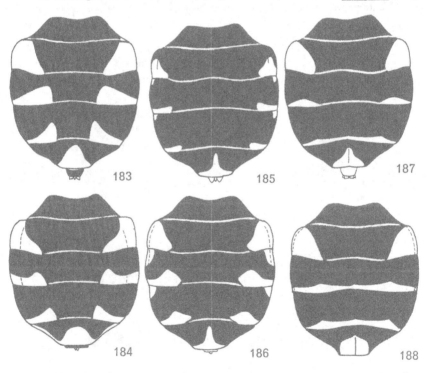

Figs. 183-188. Abdomens in dorsal view of Stratiomys. - 183: Male of S. chamaeleon (L.); 184: Female of same; 185: Male of S. furcata Fabr.; 186: Female of same; 187: Male of S. potamida Meig.; 188: Female of same.

Larvae

16. STRATIOMYS CHAMAELEON (Linné, 1758)
    Figs. 37, 43, 179, 183-184, 195-202.

Musca chamaeleon Linné, 1758: 589.

Eyes bare in both sexes. Lateral markings on abdomen conspicuous, at posterior margins of tergites, always separated. Venter mainly yellow with narrow transverse spots or stripes. Yellow occipital band well-developed in females, interrupted only at middle and female face largely yellow. - Male genitalia: Epandrium large, cerci almost oblong. Medial process of synsternite
high, with short-conical and backwardly directed spines laterally. Dististyles
with two sharply pointed tips. Aedeagal complex stout, especially in middle,
base of parameres spinulose. - Length: body 12.0-16.0 mm, wing 9.8-12.3
mm.

    The larva has been inadequately described by several earlier authors. The
diagnostic characters have been emphasized by Dušek & Rozkošný (1965). Colour ochre to greyish and black with darker longitudinal stripes. Pubescence on

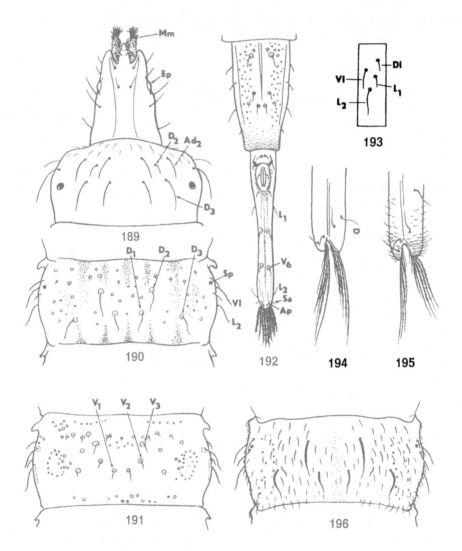

Figs.189-196. Morphology of Stratiomys-larvae. - 189: Anterior end in dorsal view of S.longicornis (Scop.); 190: Abdominal segment 3 in dorsal view of same; 191: Same in ventral view; 192: Posterior end in ventral view of same; 193: Scheme of setae on lateral wall of abdominal segment 3 of same; 194: Apex of anal segment in lateral view of same; 195: Apex of anal segment in lateral view of S.chamaeleon (L.); 196: Abdominal segment 3 in ventral view of same. Abbreviations: see p.132.

body segments rather long and strong, dark dorsally and dense laterally. Anal segment about 8 times as long as wide, with recurved bristle-like hairs in front of apical coronet. Cf2 seta on head simple. - Length: 45-58 mm, maximum width: 5.4-7.5 mm. (Several larvae and puparia from Central Europe).

Not very common, but apparently widespread in North Europe. Found in some provinces of Denmark on Jutland, Fyn and Zealand; not found in Norway but recorded from C. and S. Sweden (Sk., Hall., Ög., Vg., Sdm., Upl., Jmt.),

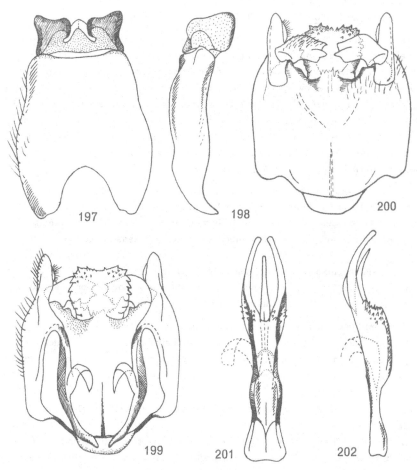

Figs. 197-202. Male genitalia of Stratiomys chamaeleon (L.). - 197: Dorsal part in dorsal view; 198: Dorsal part in lateral view; 199: Ventral part in dorsal view; 200: Ventral part in ventral view; 201: Aedeagal complex in dorsal view; 202: Aedeagal complex in lateral view.

and also in Soviet Carelia. - The greater part of Europe, through C. Asia and Siberia to Lake Baical. - June-August. - Larvae live in pools and ponds, floating on the water-surface among vegetation, and often hibernate in mud. Adults in moist situations, often on various flowers.

17. STRATIOMYS FURCATA Fabricius, 1794
     Figs. 180, 185-186, 203-207.

Stratiomys furcata Fabricius, 1794: 264.
Stratiomys panthaleon Fallén, 1817: 7.

Eyes hairy in male and bare in female. Yellow lateral markings on abdomen small and narrow in male and large in female, usually extending along anterior angles of tergites. Venter black with yellow transverse stripes. Female occiput with only two yellow spots, frons and face black with two yellow spots above antennae. - Male genitalia: Epandrium subquadrate, cerci relatively small and oval. Synsternite conspicuously elongate, with rather low, medially projecting medial process and dististyles that are simply pointed at tips. Aedeagus long and completely smooth, somewhat swollen at base of parameres. - Length: body 10.0-18.0 mm, wing 9.5-12.8 mm.

The larva was briefly described by Lundbeck (1907) and Séguy (1926), and adequately by Dušek & Rozkošný (1965). Larva grey or brown, with darker longitudinal stripes. Fine setae on body segments almost indistinct among the rather long pubescence, but this pubescence pale and less dense than in cha-maeleon. Lateral projections more distinct on segments 2 and 3 but rather small. Anal segment very long, about 12-14 times as long as wide, bare in front of tip except for the constant setae and the apical coronet of float-hairs. - Length: 40.0-60.0 mm, maximum width: 5.0-7.0 mm.

The commonest nordic species of the genus. Known from all provinces of Denmark except W. Jutland and Bornholm; widespread in many provinces of S. and C. Sweden and Finland, in N to Nb. and Ob.; also in S. Norway and in Soviet Carelia. - Throughout Europe and Asia, eastwards to Japan. - May-August. - Larvae have been found in standing water with the puparia often floating on the water surface. Adults around the larval habitats, frequently on flowering Dauca-ceae.

Note: No specimens answering the original description were found in Fabricius' collection in ZMC. Under the name of panthaleon there are 3 males and 8 females in Fallén's collection in NRS, several of them being labelled "Str. panthaleon, furcata Fabr." These specimens all belong to furcata and apparently represent a series of Fallén's syntypes.

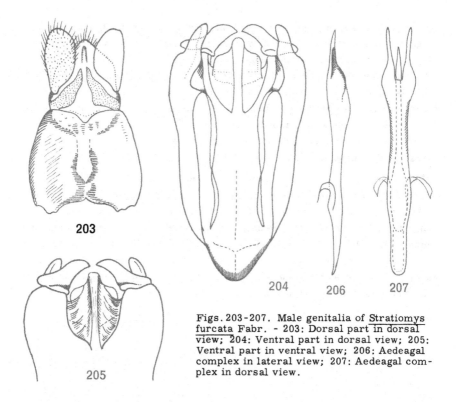

203

204 **206** **207**

205

Figs. 203-207. Male genitalia of Stratiomys furcata Fabr. - 203: Dorsal part in dorsal view; 204: Ventral part in dorsal view; 205: Ventral part in ventral view; 206: Aedeagal complex in lateral view; 207: Aedeagal complex in dorsal view.

18. STRATIOMYS LONGICORNIS (Scopoli, 1763)
Figs. 44, 181, 189-194, 208-211.

Hirtea longicornis Scopoli, 1763: 367.
Stratiomys strigata Fabricius, 1781: 417.
Stratiomys thoracica Fabricius, 1805: 79.

Eyes densely hairy in both sexes. Abdomen entirely black, very seldom with small and indistinct lateral markings developed. Abdominal pubescence some-times denser and paler on the site of the reduced lateral markings. Venter black with yellow posterior margins to sternites. Female face with yellow spots ex-tending from eye-margin to middle above antennae; occiput with two contrasting yellow spots. - Male genitalia: Epandrium simple and small, cerci almost tri-angular. Synsternite remarkably elongated, with slender inner posterolateral projections. Dististyles sharply pointed apically. Aedeagal complex consider-ably swollen at base of parameres. - Length: body 8.0-14.0 mm, wing 8.0-11.5 mm.

The larva was mentioned by Séguy (1926) and Goidanich (1939), and the main diagnostic characters were given by Brindle (1964a). Brown, with more or less distinct darker spots arranged longitudinally. Head often bluish-grey, with black spots at bases of setae. With distinct circular pale spots at bases of setae on all body segments. The spine-shaped and posteriorly directed protuberances at anterior angles of abdominal segments 1-4 well-developed. Anal segment about 8 times as long as wide, smooth in front of apical coronet except for the constant setae. - Length: 38.0-42.0 mm, maximum width: 5.0-5.8 mm. (Several larvae from Central Europe).

Rather common all over Denmark (not found in NWJ, WJ, NWZ and B at present), but rarely in S. Sweden (Lund, Alnarp and Sandby in Sk., according to Zetterstedt, 1849; also taken in Gtl. by Boheman). - A Palaearctic species known from North Africa to S. Sweden and NW USSR, and eastwards to Iran and China. - May-July. - Larvae in standing water among vegetation, and also in saline pools near sea-shore and in salt marshes; often overwintering in moist earth, under stones, etc. Adults on ground vegetation and flowers in the littoral zone and around the larval habitats.

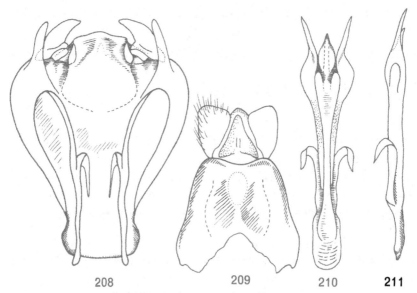

208   209   210   211

Figs.208-211. Male genitalia of Stratiomys longicornis (Scop.). - 208: Ventral part in dorsal view; 209: Dorsal part in dorsal view; 210: Aedeagal complex in dorsal view; 211: Aedeagal complex in lateral view.

Note: As regards strigata, there are 3 fragments (virtually remnants of the thorax) in Fabricius' collection (ZMC); one of these bears the label "strigata", and the other two are pinned nearby; all belong to longicornis. The type-material of thoracica is represented in Fabricius' collection (ZMC) by 2 relatively well-preserved females labelled "thoracica" and "Type", both belonging to longicornis as well.

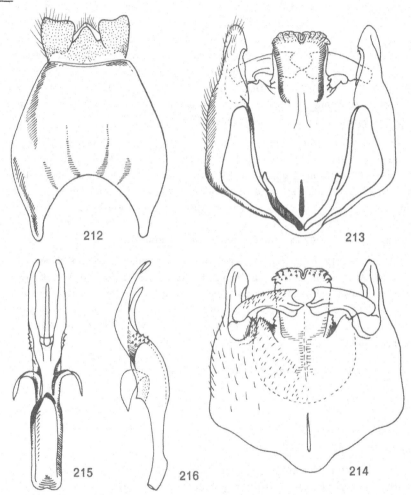

Figs. 212-216. Male genitalia of Stratiomys potamida Meig. - 212: Dorsal part in dorsal view; 213: Ventral part in dorsal view; 214: Ventral part in ventral view; 215: Aedeagal complex in dorsal view; 216: Aedeagal complex in lateral view.

19. STRATIOMYS POTAMIDA Meigen, 1822

Figs. 178, 182, 187-188, 212-216.

Stratiomys potamida Meigen, 1822: 136.

Eyes bare in both sexes. The male in particular resembles chamaeleon in general appearance. However, the yellow lateral markings on abdomen are connected medially on tergite 4 in the male and on tergite 3 and 4 in the female. Venter mainly yellow. Female frons completely black and face yellow except for narrow median stripe and margin of oral opening. Post-ocular band in females broadly yellow, and is interrupted medially so that vertex and upper occiput are black. - Male genitalia: Medial process of synsternite high, spinose posterolaterally, with a fine median incision. Dististyles with two pointed tips. Aedeagal complex formed as in chamaeleon, including spinules at base of parameres. - Length: body 13.0-16.0 mm, wing 10.0-12.6 mm.

The larva has been described by Brindle (1964a) and Dušek & Rozkošný (1965). Colour dark-grey or blackish, paler pattern almost indistinct. Larvae very similar to those of chamaeleon, but in addition to the branched and pubescent Lb2 setae the Cf2 setae are usually also bush-like. Other setae and pubescence as in chamaeleon, including short backwardly directed bristle-like hairs on anal segment in front of apical coronet. Anal segment about 9-10 times as long as wide. - Length: 35.0-48.0, maximum width: 4.8-5.2 mm.

Occurring only in Denmark and southernmost Sweden. It seems to be rather widespread in Denmark (W., E. and S.Jutland, NEZ, LFM and B), but in Sweden is only known from some localities in Sk. - The greater part of Europe but apparently absent in Fennoscandia except S.Sweden. - June-August. - Larvae live in the littoral zone of standing water and in marshes near streams. Adults on low herbage and flowers.

Genus Odontomyia Meigen, 1803

Eulalia Meigen, 1800, Nouv.class., 21 (suppressed by I.C.Z.N.).
Odontomyia Meigen, 1803, Mag.Insektenkd., 2: 265.
Trichacrostylia Enderlein, 1914, Zool.Anz., 43: 607.
Type-species: Musca hydroleon Linné, 1758.

Usually smaller species than those of Stratiomys. Third antennal segment consisting of 6 flagellomeres, first segment at most twice as long as the second. Of the M-veins arising from the discal cell, M3 is considerably reduced or absent, and M1 is remarkably shortened and fine in some species. Aedeagal complex always short and rather wide, with short compact basal part.

Larvae usually with a shorter anal segment than in Stratiomys, at most 4 times as long as wide. Penultimate, and sometimes also antepenultimate, abdominal segment with 1-2 pairs of ventral hooks on posterior margin in some species. Pubescence of varying lengths and densities developed in addition to the main setae; in some cases the hairs forming the pubescence short and dilated, almost scale-shaped. V3 setae on head situated more laterally, and consequently above and before V2. V3 setae simple, at most only pubescent.

Only 6 species are known from North Europe.

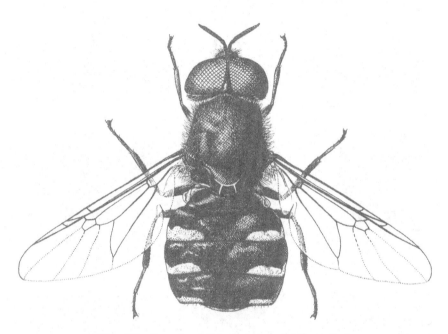

Fig. 217. Female of Odontomyia microleon (L.).

Key to species of Odontomyia

Adults

1   Basal part of the reduced M3 always distinct 2
-      M3 entirely absent (Fig. 217) 4
2 (1) Larger species (about 12-15 mm); lateral markings on ab-
       domen orange (Figs. 227-228) .... 24. ornata (Meig.)

- Smaller species (mostly 8-10 mm); lateral markings on
  abdomen green or pale yellow (Figs. 223, 226)...................... 3
3 (2) Third antennal segment relatively short and mostly yellow
  (Fig. 218); median black spots on abdomen not remarkably
  extended at anterior margins of tergites (Fig. 223).. 20. angulata (Panz.)
- Third antennal segment relatively long and brown (Fig. 219);

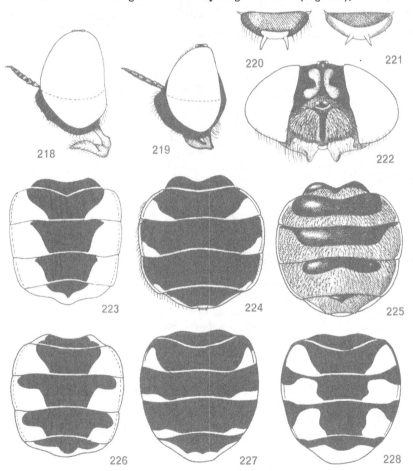

Figs. 218-228. Morphology of adults of Odontomyia. - 218: Male head in lateral
view of O. angulata (Panz.); 219: Male head in lateral view of O. hydroleon (L.);
220-221: Male and female scutellum of same; 222: Female head in frontal view
of O. ornata (Meig.); 223: Male abdomen of O. angulata (Panz.); 224-225: Male
and female abdomen of O. argentata (Fabr.); 226: Female abdomen of O. hydroleon (L.); 227-228: Female and male abdomen of O. ornata (Meig.).

median black spots on abdomen remarkably extended   at
anterior margins of tergites (Fig. 226)............. 22. hydroleon (L.)

4 (1)   First antennal segment hardly longer than the second; dorsum of abdomen shining black, venter yellow ...... 25. tigrina (Fabr.)

-   First antennal segment about twice as long as the second;
abdomen of a different colour 5

5 (4)   Dorsum of abdomen shining black, with narrow, stripe-like
lateral markings (Fig. 217) .... 23. microleon (L.)

-   Dorsum of abdomen dull, mostly covered with silvery
(male), or transversely striped with golden-grey (female),
adpressed hairs; male abdomen with triangular yellowish
lateral markings (Fig. 224).................... 21. argentata (Fabr.)

Larvae

1   Ventral hooks on posterior margin of penultimate abdominal segment present (Fig. 231) 2

-   Ventral hooks on posterior margin of penultimate segment
absent (Figs. 230, 232) 3

2 (1)   Ventral hooks only on abdominal segment 7....... 20. angulata (Panz.)

-   Ventral hooks on abdominal segments 6 and 7 (Fig. 231)..
.... 24. ornata (Meig.)

3 (1)   Anal segment almost oblong, only projecting a little posteromedially (Fig. 230) .... 21. argentata (Fabr.)

-   Anal segment almost conical, distinctly produced posteromedially (Fig. 232) .... 25. tigrina (Fabr.)

20. ODONTOMYIA ANGULATA (Panzer, 1798)
Figs. 218, 223, 239-242.

Stratiomys angulata Panzer, 1798: 19.
Stratiomys ruficornis Zetterstedt, 1842: 139.

Closely related to hydroleon and often confused with it. Head relatively wide,
especially in the male. Antennae shorter, yellow to yellowish-brown, at most
darkened at tip. Face less projecting than in hydroleon. Black spot on scutellum variable, usually not as large as in hydroleon, occupying basal half or even
less of scutellum. Black patches on abdominal tergites not strikingly extended
anterolaterally. - Male genitalia: Medial process of synsternite relatively narrow and high, dististyles almost rectangular, oblong. Aedeagal complex slight-

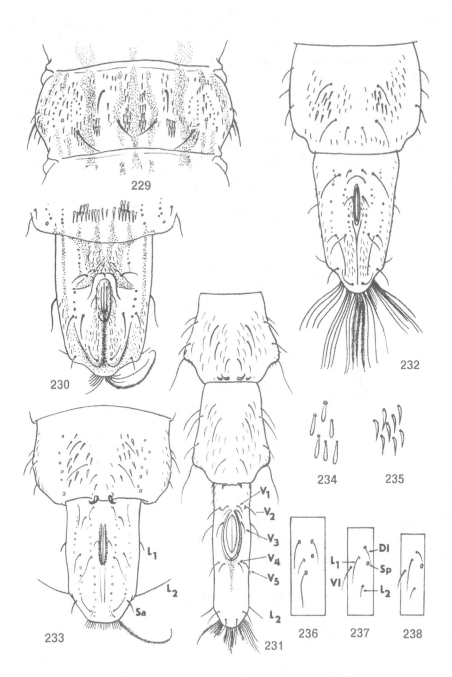

ly tapered apically, with short basal aedeagal apodeme. - Length: body 8.0-12.0 mm, wing 7.8-10.2 mm.

The larva was briefly described by Lundbeck (1907) and figured by Goidanich (1939). Brown, with some indistinct longitudinal stripes. A pair of curved ventral hooks on posterior margin of penultimate segment. First six abdominal segments each with a transverse ring of similar but smaller spines on posteroventral margins. Length: about 20.0 mm.

Not very common, but well-known from some provinces of Denmark (EJ, F, NEZ, SZ, LFM) and in S. and E. Sweden (Sk., Öl., Gtl., G. Sand.). In Finland only on Åland Islands. - Europe, S. Siberia and China. - June-August. - The larva was found in a lake, and adults on waterside vegetation.

21. ODONTOMYIA ARGENTATA (Fabricius, 1794)
Figs. 224-225, 229-230, 243-246.

Stratiomys argentata Fabricius, 1794: 266.

A medium-sized dark species, with silvery or more golden short, adpressed pubescence on abdomen. First antennal segment twice as long as the second. Scutellar spines short, indistinct. M1 remarkably reduced, R4 absent. Dark-orange lateral markings on male abdomen rather distinct. - Male genitalia: Dististyles with a blunt inner basal lobe and tapering tip. Medial process of synsternite low and simple. Aedeagal complex simple, slightly curved ventrally in apical part. - Length: body 8.0-10.0 mm, wing 7.3-8.5 mm.

The larva resembles that of tigrina. No ventral hooks developed on abdominal segments. Pubescence consisting of silvery hairs that are dilated basally, long and adpressed. Ventral pubescence arranged in a broad median longitudinal stripe and two narrower lateral stripes. The longest hairs in front of the last third of the segment form a rather dense, irregular transverse row. Anal segment almost oblong, about 1.5 times as long as wide, little projecting posteromedially. Puparium light-yellow to brown with dark-brown pattern. - Length: about 18.0 mm, maximum width: 3.0 mm. (2 puparial exuviae without anterior part in ZIL).

Figs. 229-238. Morphology of larvae of Odontomyia and Oplodontha. - 229: Abdominal segment 3 in dorsal view of Odontomyia argentata (Fabr.); 230: Anal segment in ventral view of same; 231: Posterior end in ventral view of Odontomyia ornata (Meig.); 232: Same in Odontomyia tigrina (Fabr.); 233: Same in Oplodontha viridula (Fabr.); 234-235: Dorsal and ventral hairs on abdominal segments of same species; 236-238: Scheme of setae on lateral wall of abdominal segment 3 in: 236: Odontomyia ornata (Meig.); 237: O. tigrina (Fabr.) and 238: Oplodontha viridula (Fabr.). Abbreviations: see p. 132.

A relatively rare species. In Denmark taken on Zealand, Lolland and Falster; in Sweden known from Sk., Ög., Vg., Upl. and Vrm.; only 2 females known from Finland (Helsingfors, Pasila). - Europe except for the S.parts, Siberia, Mongolia. - April-June. - Larvae were found in flood-refuse and in a moist rotting alder tree (Alnus). Males are reported to fly rapidly, sometimes hovering in small swarms. Females have been colledted on flowering Salix.

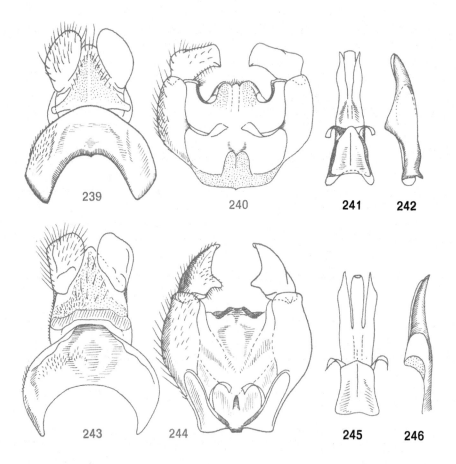

Figs.239-246. Male genitalia of 239-242: Odontomyia angulata (Panz.) and 243-246: O.argentata (Fabr.). - 239 and 243: Dorsal part in dorsal view; 240 and 244: Ventral part in dorsal view; 241 and 245: Aedeagal complex in dorsal view; 242 and 246: Aedeagal complex in lateral view.

22. ODONTOMYIA HYDROLEON (Linné, 1758)
Figs. 219-221, 226, 247-250.

Musca hydroleon Linné, 1758: 589.

Differing from the closely-related angulata by the smaller head, and longer and mostly dark brown antennae. Face more projecting than in angulata. Scutellum predominantly black, often only yellow between spines. Median black patches on abdominal tergites extended laterally along anterior margin. - Male genitalia: Medial process of synsternite wide and low. Dististyles rounded on outer surface. Aedeagal complex slightly narrowed in middle, aedeagal apodeme virtually absent. - Length: body 8.0-12.0, wing 6.4-8.5 mm.

The larva is not known.

With almost the same distribution as angulata but, in addition to Denmark, extending further north in Sweden (Sk., Ög., Vg. and Jmt.), and also in S. Norway (Ak), Aland Islands of Finland and Soviet Carelia. - The whole of Europe, from Italy and Bulgaria to C. Fennoscandia, Siberia and China. - June-July. - On ground-vegetation by standing water.

23. ODONTOMYIA MICROLEON (Linné, 1758)
Figs. 217, 251-254.

Musca microleon Linné, 1758: 589.

Medium sized species with rounded shining abdomen. Eyes bare, first antennal segment twice as long as the second. Scutellum dark, with long yellow spines. Wings with M1 reduced and M3 absent. Pale yellow or white lateral markings on abdomen very narrow, stripe-like, on posterior margins of tergites. Venter pale yellow with a pair of small black spots at middle of sternites 3 and 4. - Male genitalia: Medial process of synsternite low, weakly bilobed. Dististyles nearly triangular. Aedeagal complex almost parallel-sided in dorsal view and straight in lateral view. - Length: body 8.0-11.0 mm, wing 6.0-8.2 mm.

The larva has not been described.

Not very common but apparently widespread in the area treated here. In Denmark found in Jutland and NE Zealand, and in Sweden known from S. and C. provinces (Sk., Öl., Ög., Vg. and Upl.); also taken in S. Norway (Ø, Ak) and in several provinces of eastern Fennoscandia including Soviet Carelia. - N. part of the Palaearctic Region (C. and N. Europe, Siberia, Mongolia, China). - May-June. - Usually in rather moist situations.

24. ODONTOMYIA ORNATA (Meigen, 1822)

Figs. 38, 45, 222, 227-228, 231, 236, 255-258.

Stratiomys ornata Meigen, 1822: 144.

The largest species of the genus. First antennal segment only slightly longer than the second. Female head mainly black, with a pair of almost semicircular, longitudinally placed yellow spots. 3 M-veins arising from the discal cell, though M3 considerably reduced. Abdomen flattened, black, with 3 pairs of large orange lateral markings of almost equal size except for the single apical spot. - Male genitalia: Epandrium large, proctiger conical with oval cerci. Medial process

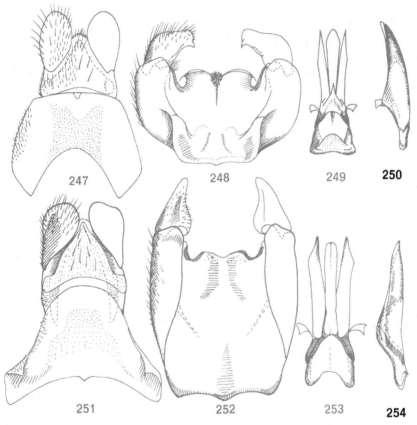

Figs. 247-254. Male genitalia of 247-250: Odontomyia hydroleon (L.) and 251-254: O. microleon (L.). - 247 and 251: Dorsal part in dorsal view; 248 and 252: Ventral part in dorsal view; 249 and 253: Aedeagal complex in dorsal view; 250 and 254: Aedeagal complex in lateral view.

of synsternite low and flat. Dististyles almost straight, triangular. Aedeagal complex simple, parameres pointed. - Length: body 12.0-15.2 mm, wing 9.8-12.2 mm.

Larva. The diagnostic characters were mentioned by Lundbeck (1907) and Séguy (1926), and also by Brindle (1964a), and a detailed description has been published by Dušek (1961). Colour reddish-brown or dark brown with a longitudinal pattern. V3 on head pubescent and long; ventral setae on abdominal segments almost indistinct among the long and pale pubescence. 8 pairs of rudimentary spiracles on metathorax and first 7 abdominal segments. Anal segment extremely long compared with other larvae of the genus, about 3.5-4 times as long as wide. 1-2 pairs of distinct ventral hooks on posterior margins of abdo-

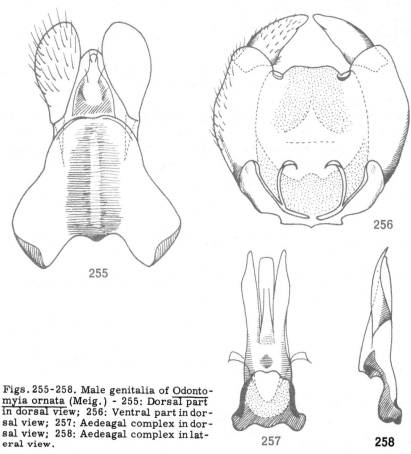

Figs. 255-258. Male genitalia of Odontomyia ornata (Meig.) - 255: Dorsal part in dorsal view; 256: Ventral part in dorsal view; 257: Aedeagal complex in dorsal view; 258: Aedeagal complex in lateral view.

minal segments 6 and 7. - Length: 31.0-38.0 mm, maximum width: 5.0-5.2 mm. (Several larvae and puparia from Central Europe).

Known from most parts of Denmark (Jutland, Zealand, Fyn, Langeland, Falster, Lolland), but in Sweden reported only from Blekinge and Östergötland (Zetterstedt, 1842, 1849, 1855). Not known from Norway or eastern Fennoscandia. - The whole of Europe, N. to C. Sweden, and also Asia Minor and Siberia. - May-July. - Larvae in shallow water, in the littoral zone of ponds, and puparia often in flood-refuse or floating on the water-surface. Adults on grass, shrubs and flowers near the larval habitat.

25. ODONTOMYIA TIGRINA (Fabricius, 1775)
Figs. 232, 237, 259-262.

Stratiomys tigrina Fabricius, 1775: 760.
Stratiomys nigrita Fallén, 1817: 9.

A predominantly black species with yellow or greenish tibiae, tarsi and venter of abdomen. Eyes hairy in both sexes, though hairs very short in female. First antennal segment rather longer than the second. Female abdomen occasionally with small orange spots at posterior corners of tergites 2 and 3. - Male genitalia: Paraprocts of proctiger narrow and semicircular. Medial process of synsternite of medium height and flat. Dististyles slightly curved inwards, pointed at tips. Aedeagal complex with rather long and well-separated parameres. - Length: body 7.5-10.5 mm, wing 6.0-6.8 mm.

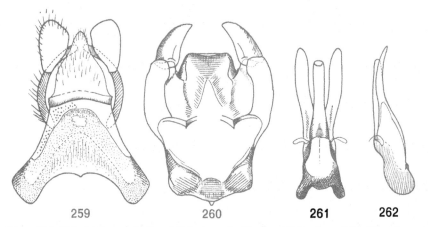

259   260   261   262

Figs. 259-262. Male genitalia of Odontomyia tigrina (Fabr.). - 259: Dorsal part in dorsal view; 260: Ventral part in dorsal view; 261: Aedeagal complex in dorsal view; 262: Aedeagal complex in lateral view.

The larva was mentioned by Brindle (1964a) and was adequately described by Dušek & Rozkošný (1965). Yellowish to brown with narrow pale longitudinal stripes, 2 pairs of which are usually more distinct medially on abdominal segments. Setae on head developed normally, V3 simple and long. Dorsal and ventral pubescence on body segments not very dense, consisting of whitish hairs that are dilated basally and adpressed on the dorsal side. Ventral hooks absent. Anal segment 1.5-2.0 times as long as wide, distinctly produced posteromedially. - Length: 16.5-22.0 mm, maximum width: 3.0-4.3 mm.

Common in Denmark (all provinces except for WJ), S. and C. Sweden (Sk., Öl., Ög., Vg., Nrk. and Upl.). Not known from Norway and eastern Fennoscandia. - In Europe from Jugoslavia and Bulgaria to C. Sweden, Siberia. - May-July. - Larvae occur in shallow pools, at the margins of ponds and in marshes among decaying vegetable matter. Adults often on flowers of Daucaceae.

Genus Oplodontha Rondani, 1863

Oplodontha Rondani, 1863, Arch. zool. Mod., 3: 78.
Hoplodonta, emend.
Type-species: Stratiomys viridula Fabricius, 1775.

Similar to Odontomyia but generally smaller, with green or yellowish ground-colour to abdomen and a variable black pattern. Third antennal segment consisting of 6 flagellomeres. Wings with Rs simple (or R1 united with R2+3), discal cell very small, M1 considerably reduced and M3 absent. Male genitalia resembling those of Odontomyia spp., with medial process of synsternite conical.

The larvae with conspicuous V3 on head which are large, bush-like, and formed by numerous pubescent branches. Dorsal pubescence on abdominal segments dense, consisting of scale-like, apically dilated short hairs. Penultimate segment with strong ventral hooks. Anal segment relatively short, about 1.3-1.8 times as long as wide. The genus is widely distributed in the tropics but is represented by only 1 species in Europe.

26. OPLODONTHA VIRIDULA (Fabricius, 1775)
Figs. 46, 233-235, 238, 263-270.

Stratiomys viridula Fabricius, 1775: 760.

A small species with reduced wing-venation (see above). Eyes bare in both sexes. Scutellum semi-elliptical, with 2 small but distinct spines. Ground

colour of abdomen varying from light-green to yellowish-brown, venter always
without any black spots. Median dorsal patches forming a narrower (males) or
wider (females) mid-dorsal black stripe with irregular edges. Many of the va-
rieties with a reduced dorsal pattern on abdomen were described as distinct
species by earlier authors. - Male genitalia: Medial process of synsternite
simple and conical, dististyles with small apical lobe. Aedeagal complex tapering apically, and parameres narrow and pointed in dorsal view, slightly curved
dorsally. - Length: body 6.0-9.0 mm, wing 5.4-6.8 mm.

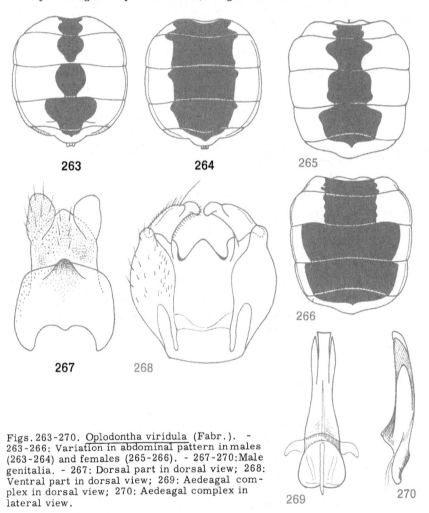

Figs. 263-270. Oplodontha viridula (Fabr.). -
263-266: Variation in abdominal pattern in males
(263-264) and females (265-266). - 267-270: Male
genitalia. - 267: Dorsal part in dorsal view; 268:
Ventral part in dorsal view; 269: Aedeagal complex in dorsal view; 270: Aedeagal complex in
lateral view.

The larva was briefly described by Séguy (1926) and in greater detail by Du-šek (1961). Dark brown to grey, with a conspicuous median stripe on abdominal segments and short oblong lateral stripes on each side. 8 pairs of rudimentary spiracles distinct in addition to larger anterior spiracles. Other characters as given above. - Length: 12.8-17.0 mm, maximum width: 2.4-3.1 mm. (Several larvae and puparia from Central Europe).

Common in the nordic area. Frequent in all provinces of Denmark and also in Fennoscandia, from S to NT in Norway, Jmt. in Sweden and ObS in Finland; also in Soviet Carelia. - A Palaearctic species distributed from North Africa and Europe to China and Kamchatka. - May-September. - Larvae in marshes, pools and at the margins of ponds. Adults on ground vegetation around the larval habitats.

Subfamily Clitellariinae

The subfamily consists of a number of very diverse genera. Antennae elongate or relatively short. Third antennal segment consisting of 6 flagellomeres, the last of which is tapering and forms a more or less distinct apical style. Some species of Oxycera have two very slender apical flagellomeres in the form of a 2-segmented terminal or subterminal arista. Scutellum unarmed or with a pair of spines. Wings with M4 arising directly from the discal cell. Male genitalia are characterized by the different medial process of synsternite and the form of the aedeagal complex. The larvae are terrestrial (Clitellaria), or aquatic or semiaquatic (Oxycera, Nemotelus) in which case the anal segment is with a coronet of float-hairs or is bilobed posteriorly. Antennae situated far behind anterior end of the head-capsule. Rudimentary pupal spiracles were found on puparia of Nemotelus.

In Europe there are about 80 species in 7 genera, of which 15 species in 3 genera belong to the nordic fauna.

Genus Nemotelus Geoffroy, 1762

Nemotelus Geoffroy, 1762, Hist. abrég. Ins., 2: 542.
Type-species: Musca pantherina Linné, 1758.

Usually rather small species, mostly black and yellow or white, immediately distinguishable by the conically produced face that forms a rostrum. The so-called rostral index (maximum eye-length: length of rostrum from tip to eye-margin) is of some diagnostic importance. Antennae spindle-shaped, third segment consisting of 4 larger flagellomeres and an apical, 2-segmented style. Eyes large and touching in males, small and widely separated in females. Scutellum always unarmed. Abdomen always broader than thorax, often almost round, quite differently coloured in the two sexes. The male genitalia with a number of important diagnostic characters, especially shape of synsternite, its medial process, dististyles and aedeagal complex.

Larvae with the posterior spiracular opening situated dorsally and surrounded by short float-hairs. Anal segment deeply notched posteriorly in middle and consequently terminating in two conical lobes. Puparia with very short pupal spiracles on abdominal segments 1-6.

Only 4 rather common species in North Europe.

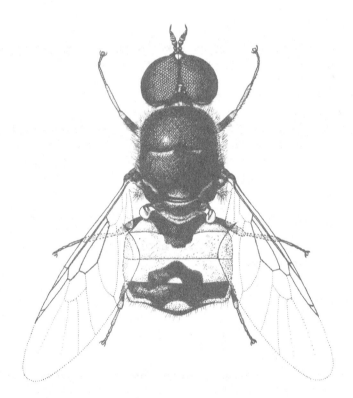

Fig. 271. Male of <u>Nemotelus notatus</u> Zett.

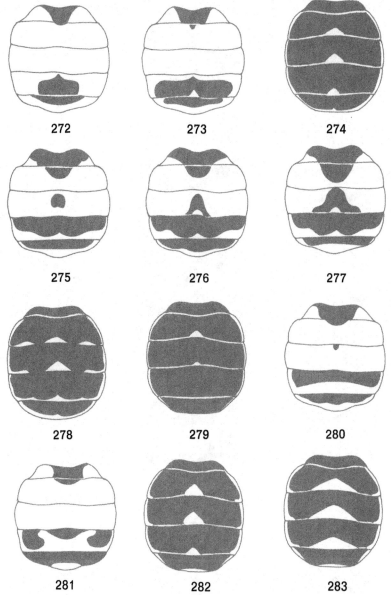

Figs. 272-283. Abdomens of Nemotelus. 272-273: Males of N. pantherinus (L.);
274: Female of same; 275-277: Males of N. notatus Zett.; 278-279: Females of
same; 280-281: Males of N. uliginosus (L.); 282-283: Females of same.

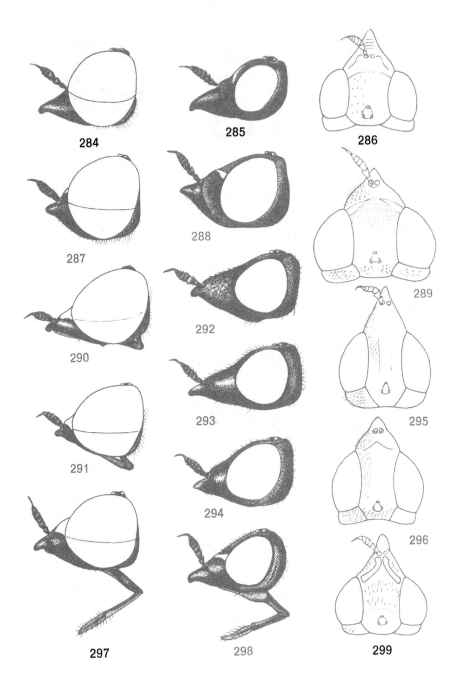

284

285

286

287

288

289

290

292

291

293

295

297

294

296

298

299

Key to species of *Nemotelus*

Adults

1   Body shining black, without any white spots on abdomen;
    vein R4+5 simple .... 27. nigrinus Fall.
-   Body black and white, abdomen predominantly white or pale
    yellow or black, at least with white spots; vein R4+5 nor-
    mally branched 2
2 (1) Venter white except for a more or less distinct basal spot;
    rostrum long and completely black in females (Figs. 292-
    294) .... 29. pantherinus (L.)
-   Venter predominantly black, at most whitish medially; fe-
    male rostrum with a pair of light spots (Figs. 289, 299).............. 3
3 (2) Stout species, with relatively short rostrum in both sexes
    (Figs. 287-289); male abdomen usually with a central spot
    on tergite 3 (Figs. 275-277) .... 28. notatus Zett.
-   Slender species, with rather long rostrum in both sexes
    (Figs. 297-299); male abdomen without a central spot on
    segment 3, at most with small anteromedian spot (Figs.
    280-281) .... 30. uliginosus (L.)

Larvae (adapted from Brindle, 1964b)

1   V3 setae on abdominal segments about 1/4 as long as V2
    (Fig. 301); apical lobes of anal segment more conical (Fig.
    300) .... 29. pantherinus (L.)
-   V3 setae on abdominal segments about 1/2 as long as V2;
    apical lobes of anal segment more cylindrical (Fig. 302) 2
2 (1) Ventral lobes constricted near base so that their greatest
    width is distal of the base; dorsum more uniformly colour-
    ed, with short, darker, curved stripes .... 30. uliginosus (L.)
-   Ventral lobes widened at base (Fig. 302); dorsum with a
    prominent grey median stripe, developed into annular mark-
    ings posteriorly, bordered with dark brown .... 28. notatus Zett.

Figs. 284-299. Heads of Nemotelus. - 284: Male head in lateral view of N. nigrinus Fall.; 285: Female head in lateral view of same; 286: Female head in
dorsal view of same; 287: Male head in lateral view of N. notatus Zett.; 288:
Female head in lateral view of same; 289: Female head in dorsal view of same;
290-291: Male heads in lateral view of N. pantherinus (L.); 292-294: Female
heads in lateral view of same; 295-296: Female heads in dorsal view of same;
297: Male head in lateral view of N. uliginosus (L.); 298: Female head in lateral view of same; 299: Female head in dorsal view of same.

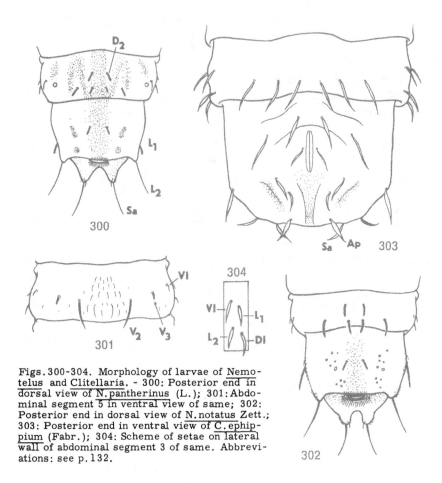

Figs. 300-304. Morphology of larvae of Nemotelus and Clitellaria. - 300: Posterior end in dorsal view of N. pantherinus (L.); 301: Abdominal segment 5 in ventral view of same; 302: Posterior end in dorsal view of N. notatus Zett.; 303: Posterior end in ventral view of C. ephippium (Fabr.); 304: Scheme of setae on lateral wall of abdominal segment 3 of same. Abbreviations: see p. 132.

27. NEMOTELUS NIGRINUS Fallén, 1817
   Figs. 284-286, 305-308.

Nemotelus nigrinus Fallén, 1817: 6.

Black species, only tips of femora and parts of tibiae and tarsi white. Antennae inserted relatively far from tip of rostrum. Vein R4+5 simple, unbranched; halteres white. - Male genitalia: Synsternite with slender and pointed processes projecting from inner posterior margin of the original basistyles. Dorsal bridge covering aedeagal complex elongated posteriorly and forming two slender dorsal apodemes. - Length: body 2.5-4.5 mm, wing 2.8-3.5 mm.

The larva is not known.

Widespread in most provinces of Denmark and in the greater part of Fenno-scandia. Extending north to Nordland in Norway, Lapland in Sweden and Ostro-bottnia in Finland. - A Palaearctic species known from North Africa, Europe to Lapland, and Asia including Mongolia and China. - May-August. - Adults on ground-vegetation in swamps and on moors.

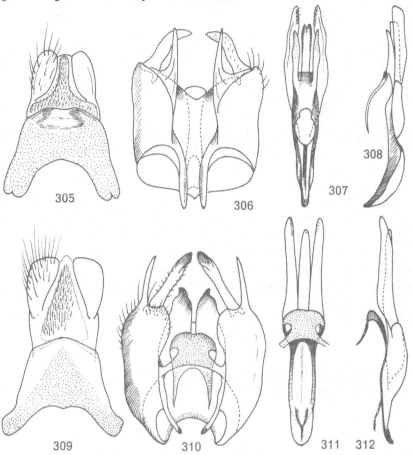

Figs. 305-312. Male genitalia of 305-308: Nemotelus nigrinus Fall. and 309-312: N. notatus (Zett.). - 305 and 309: Dorsal part in dorsal view; 306 and 310: Ventral part in dorsal view; 307 and 311: Aedeagal complex in dorsal view; 308 and 312: Aedeagal complex in lateral view.

28. NEMOTELUS NOTATUS Zetterstedt, 1842

Figs. 271, 275-279, 287-289, 302, 309-312.

Nemotelus notatus Zetterstedt, 1842: 148.

Rostral index about 3.3-3.6 in males and 1.3-1.5 in females. Black pattern on male abdomen with a variable central spot on tergite 3, and a semicircular spot on segments 1 and 2 also different. Venter black with larger yellowish patches on sternites 2 and 3. Female frons with two narrow yellow spots extending from eye-margin to middle of frons, medially tapered and separated. Light pattern on black female abdomen consisting of median triangular spots on segments 2-4, narrow lateral markings on the same segments, and a narrow margin to the whole abdomen. - Male genitalia: Synsternite with long and pointed lateral process, narrow and almost straight dististyles. Medial process bilobed, each lobe pointed at tip. Aedeagal complex simple, with well separated parameres. - Length: body 4.5-6.5 mm, wing 3.6-4.5 mm.

The larva was described briefly by Lenz (1923) and in greater detail by Brindle (1964b). Colour brownish with a broad light longitudinal stripe bordered by more or less distinct darker irregular oblique stripes. V3 on abdominal segments half as long as V2. Median posterior notch on anal segment low and rounded. L1 of anal segment rather long and inserted near middle of lateral margin. - Length: 6.0-9.0, maximum width: about 2.5 mm. (One puparial exuvia from Denmark).

Common in Denmark where it has been taken in all provinces. Apparently rarer in Fennoscandia, and apart from several provinces in S. Sweden, known only from S. Norway (Ak) and Finland (Al, Ab). - A species with a mainly maritime distribution in W. and N. Europe; records from C. and S. Europe need verification. - June-September. - Larvae live in saline pools and marshes near the coast. Adults on vegetation around the larval habitats.

29. NEMOTELUS PANTHERINUS (Linné, 1758)
   Figs. 272-274, 290-296, 300-301, 313-316.

Musca pantherina Linné, 1758: 590.
Nemotelus marginellus Fallén, 1817: 5.
Nemotelus fraternus Loew, 1846b: 446.
Nemotelus gracilis Loew, 1846b: 447.

Rostral index very variable, 2.7-4.3 in male and 1.2-1.7 in female. Frons usually black in both sexes, though rarely white including rostrum in some males. Black pattern on male abdomen confined to a basal spot and transverse patches at posterior margin of tergite 4 and anterior margin of tergite 5, but both patches may be considerably reduced. Venter entirely white except for a small basal spot. Female abdomen white marginate, with triangular median spots on tergites 2-4. - Male genitalia: Synsternite with the usual posterolateral projections and a conspicuously narrow, bipartite medial process. Aedeagal complex slender, with ventral spinules on apical half. Parameres only distinct apically. - Length: body 4.0-5.5 mm, wing 3.3-4.5 mm.

The larva was mentioned by Lundbeck (1907), Lenz (1923) and Séguy (1926), and an adequate description was published by Dušek & Rozkošný (1967). Yellowish to dark-brown with darker pattern. Dark longitudinal stripe usually distinct

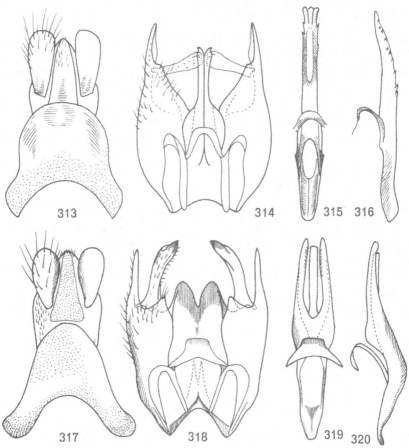

Figs. 313-320. Male genitalia of 313-316: Nemotelus pantherinus (L.) and 317-320: N. uliginosus (L.). - 313 and 317: Dorsal part in dorsal view; 314 and 318: Ventral part in dorsal view; 315 and 319: Aedeagal complex in dorsal view; 316 and 320: Aedeagal complex in lateral view.

on mid-line of abdominal segments. Anal segment deeply notched posteriorly, the two apical lobes conical. Setae on body segments mostly short, dilated and pubescent, only L2 and Sa on anal segment long. V3 setae on abdominal segments only a quarter as long as V2, V1 indistinct among the rather dense pubescence covering middle third of ventral side. Very small pupal spiracles distinct on dorsolateral part of abdominal segments 2-5. - Length: 7.0-10.5 mm, maximum width: 1.7-2.0 mm. (Several larvae and puparia from Central Europe).

The species is known from several provinces in Denmark to S. and C. Sweden. It is not known for certain from Norway or Finland, but is known from Soviet Carelia. - All over Europe to C. Sweden, also North Africa. - May-August. - Larvae in shallow standing water, in marshes, pools, etc., often together with larvae of Odontomyia and Stratiomys; adults in the same places.

Specimens with extremely short rostrum and reduced abdominal patches have been determined as fraternus or gracilis by earlier authors (cf. Lyneborg, 1960). Both these taxa are identical with pantherinus, as has been shown by examination of the original type-material (Dušek & Rozkošný, 1966, 1967).

30. NEMOTELUS ULIGINOSUS (Linné, 1767)
   Figs. 280-283, 297-299, 317-320.

Musca uliginosa Linné, 1767: 983.

Rostrum relatively long in both sexes, rostral index about 2.2-2.8 in males and 1.2-1.5 in females. Frons with white spots in both sexes. Female frontal spots in the form of a pair of narrow, medially convergent stripes. Male abdomen with black pattern consisting of a basal spot and 2 transverse black bands at anterior margins of tergites 4 and 5. Sometimes a small black anteromedian spot on tergite 3 distinct. On the other hand, the black band on segment 4 may be reduced to 3 patches. Venter completely or mainly black, at most with light patches in middle. In addition to the usual median triangles, female abdomen with white margins partly extending along posterior angles of tergites. - Male genitalia: Lobes of medial process more rounded and parameres less separated than in notatus. - Length: body 4.0-6.5 mm, wing 3.5-5.0 mm.

The larva was inadequately described by Haliday (1857) and Brindle (1964b). The diagnostic characters given by latter author (see Key to species) do not seem to be very conclusive. - Length: about 6.0 mm.

The species has been taken in most provinces of Denmark, but rather rarely in Norway (Ak, VE, NTi); it is also widespread in S. and C. Sweden, as well as in Finland and in Soviet Carelia. - C. and N. Europe, through Siberia to eastern Asia. - June-September. - Larvae live among vegetation near the surface of standing water and are often found by the sea-shore; adults in the same places.

Genus Clitellaria Meigen, 1803

Potamida Meigen, 1800, Nouv. class., 22 (Suppressed by I.C.Z.N.).
Ephippium Latreille, 1802, Hist. nat. Crust. Ins., 3: 448 (nec Bolten, 1798).
Clitellaria Meigen, 1803, Mag. Insektenkd., 2: 265.
Ephippiomyia Bezzi, 1902, Zeitschr. Syst. Hym. Dipt., 2: 191.
Type-species: Stratiomys ephippium Fabricius, 1775 (mon.).

The genus is characterized by a pair of strong lateral spines on the thorax in front of the wing-base. Only one species in Europe.

31. CLITELLARIA EPHIPPIUM (Fabricius, 1775)
Figs. 47, 303-304, 321-324.

Stratiomys ephippium Fabricius, 1775: 759.
Ephippium thoracicum Latreille, 1804: 192.

A stout and dark species. Mesonotum with short and very dense reddish-brown hairs, wings blackish infuscated. - Male genitalia: Synsternite elongate, with well-developed dorsal bridge and triangular apically pointed cerci. Aedeagal complex relatively small, aedeagus reduced, parameres tapered apically. - Length: body 12.0-14.0 mm, wing 10.2-12.5 mm.

The larva was described in adequate detail by Dušek & Rozkošný (1967). Anal segment almost oblong, with a posterodorsal transverse spiracular opening, without any apical coronet. Setae on body-segments short but conspicuous, dilated and flat with frayed margins. Dorsal setae in one transverse row, parallel or slightly convergent, ventral setae parallel or divergent. - Length: 25.0-32.0 mm, maximum width: 4.4-4.9 mm. (Several larvae from Central Europe).

Reported from Norway by Siebke (1877) but without any further data. A specimen in ZMO was taken at Fjeldotuin (1844, Esmark). In Sweden, correctly recorded from material in ZIL from Öl. and Sm. (cf. also Zetterstedt, 1850, 1855). - Distributed throughout Europe but apparently very rare in N (England, S. Norway and Sweden). - May-July. - Larvae live in nests of the ant Lasius fuliginosus, and have also been found at the roots of a walnut tree and in forest earth with decaying vegetable matter. Development probably lasts 3 years. Mature larvae leave the ant-nests and pupate nearby.

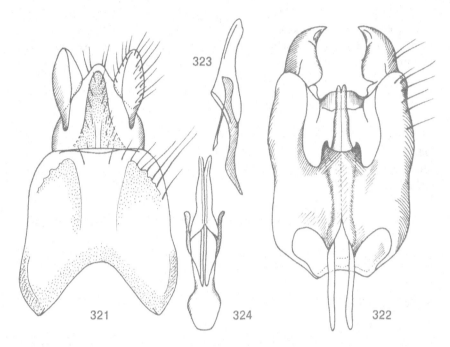

Figs. 321-324. Male genitalia of Clitellaria ephippium (Fabr.). - 321: Dorsal part in dorsal view; 322: Ventral part in dorsal view; 323: Aedeagal complex in lateral view; 324: Aedeagal complex in dorsal view.

Genus Oxycera Meigen, 1803

Hermione Meigen, 1800, Nouv.class., 22 (suppressed by I.C.Z.N.).
Oxycera Meigen, 1803, Mag.Insektenkd., 2: 265.
Type-species: Musca hypoleon Linné, 1767 = trilineata Linné, 1767.

Rather small, rarely medium-sized species, usually with contrasting yellow markings on a black ground, but some species predominantly green or yellowish with a black pattern. Third antennal segment oval to spindle-shaped, with 2-segmented apical style or subterminal arista. Scutellum always with a pair of spines. Four M-veins arising from the discal cell. Abdomen oval or almost round. Male genitalia with a simple epandrium, proctiger and elongate-oval cerci forming a dorsal part, and a compact synsternite with aedeagal complex and dististyles as a ventral part. Medial process developed as simple rounded plate or bilobed. Aedeagus normally fused with parameres in basal part. Parameres clearly distinct, rarely bipartite at tips or spinose.

Fig. 325. Male of _Oxycera trilineata_ (L.).

Larvae aquatic, with a large and conspicuous apical coronet of float-hairs. Antennae located almost dorsally between anterior end of head capsule and eye prominence. Ventral hooks on posterior margin of penultimate segment well-developed or absent. Anal segment round, or with short lobes at posterolateral angles, or with pointed lateral lobes. The shape of the sclerotized lips above and below posterior spiracular opening (dorsal and ventral spiracular plates) presents an important diagnostic character. Setae on body segments simple and smooth, plumose, gelatinous or incrustate. Abdominal segments with a transverse row of 1-5 pairs of mainly accessory anteroventral setae (Av) in addition to the normal ventral setae. The larvae live mainly in moss growing in marshes, streams and in hygropetric situations or in the mud in the littoral zone of pools. Hygropetric larvae in particular feed upon various kinds of algae, which has been proved by examination of their intestines (Schremmer, 1951).

There are more than 50 species in the Palaearctic Region. In Denmark and Fennoscandia 10 species are reliably known.

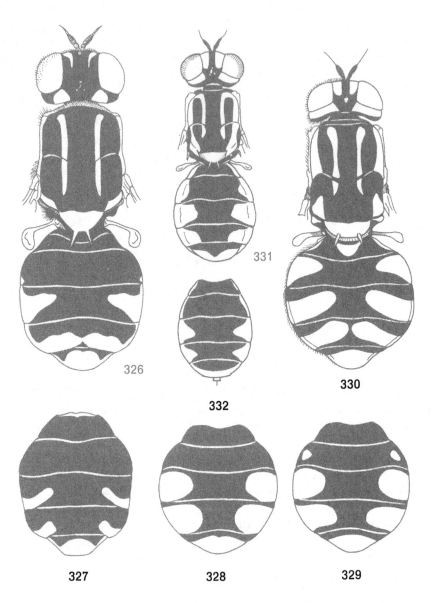

Figs. 326-332. Morphology of adults of Oxycera. - 326-327: Female of O. pardalina Meig., showing variation in pattern; 328: Male abdomen of O. dives Lw.;
329: Female abdomen of same; 330: Female of O. fallenii Stæg.; 331: Female of
O. formosa Meig.; 332: Male abdomen of same.

Key to species of Oxycera

Adults

1   Abdomen green or yellow with black pattern (Fig. 325)....
.... 41. trilineata (L.)
-    Abdomen black with contrasting yellow spots 2

2 (1) Yellow lateral markings on abdomen absent 3
-    Yellow lateral markings on abdomen present 4

3 (2) Abdomen with distinct basal and apical yellow spots; medial
process of synsternite bilobed (Fig. 373) .... 36. leonina (Panz.)
-    Abdomen with only an apical yellow spot; medial process
high and simple (Fig. 389) .... 40. terminata Meig.

4 (2) Yellow lateral markings on tergites 3 and 4 in the form of
oblique stripes (Figs. 330, 336) 5
-    Yellow lateral markings on tergites 3 and 4 usually smaller
and rounded, rarely fused medially on tergite 4 6

5 (4) Longitudinal yellow stripes on mesonotum joined to the humeral spots (Fig. 336) .... 37. meigenii Stæg.
-    Longitudinal yellow stripes on mesonotum not joined to the
humeral spots (Fig. 330) .... 33. fallenii Stæg.

6 (4) Abdominal lateral markings large and rounded, distinctly
separated at lateral margin (Figs. 328-329).............. 32. dives Lw.
-    Abdominal lateral markings smaller, differently shaped or
obviously fused on lateral margin 7

7 (6) Scutellum blackish in at least basal third; legs predominantly black in male and darkened in female; aedeagus short and
wide (Fig. 386); female mesonotum without pale longitudinal
stripes .... 39. pygmaea Fall.
-    Scutellum wholly yellow; legs seldom darkened; aedeagus
conspicuously long, but if short then medial process of synsternite bilobed; female mesonotum with yellow longitudinal stripes 8

8 (7) Male with a striking yellow patch above notopleural line;
mesonotal stripes joined to humeral spots in female (Fig.
331) .... 34. formosa Meig.
-    Male without a patch above notopleural line; female with
mesonotal stripes and humeral spots separated (Fig. 326) 9

9 (8) Male genitalia as in figs. 366-371; female abdomen with a
distinct basal spot in addition to lateral markings (Fig. 333)

103

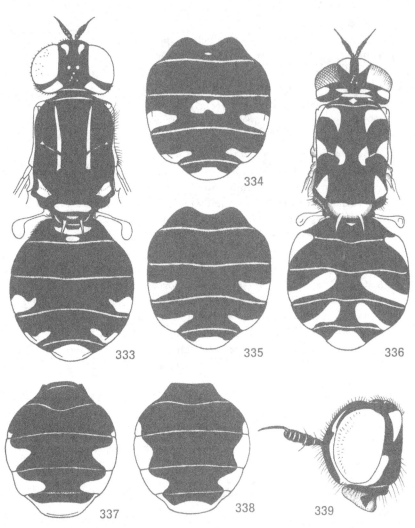

Figs. 333-339. Morphology of adults of Oxycera. - 333: Female of O. freyi (Lind.); 334: Female abdomen of holotype of O.freyi (Lind.); 335: Male abdomen of same species; 336: Female of O.meigenii Stæg.; 337: Male abdomen of O.pygmaea Fall.; 338: Female abdomen of same; 339: Female head in lateral view of O.dives Lw.

Larvae

32. OXYCERA DIVES Loew, 1845
Figs. 328-329, 339, 352-357.

Oxycera dives Loew, 1845: 15.

Eyes with dense black hairs. Female head with a complete or laterally interrupted yellow post-ocular band. Female mesonotum with 2 longitudinal stripes,
and often with triangular lateral spots in front of suture. Abdomen round, with
5 large isolated and medially rounded spots. Spots on segments 3 and 4 of almost the same shape. Small lateral markings on tergite 2 sometimes present.
- Male genitalia: Epandrium extended into oval lateral plates, medial process
of synsternite bilobed, aedeagal complex simple and tapering basally. - Length:
body 6.0-8.0 mm, wing 6.8-7.6 mm.

The larva is not known.

Only two female specimens have been taken in eastern Fennoscandia, according to the material examined. One from Ponoj, Kola Peninsula, Lr, (leg. Frey)

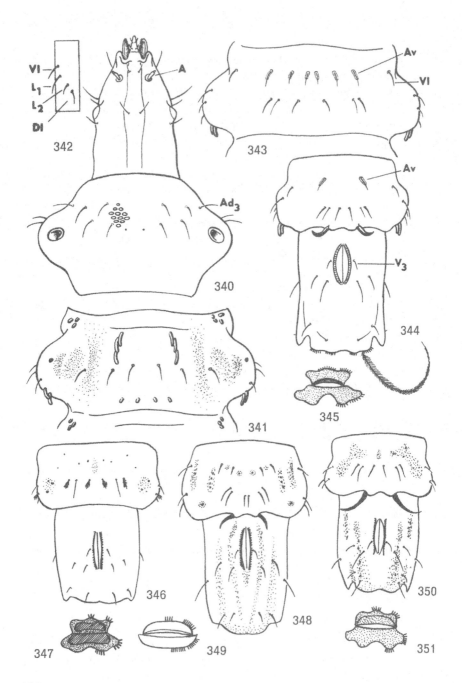

342

343

340

341

344

345

346

347

348

349

350

351

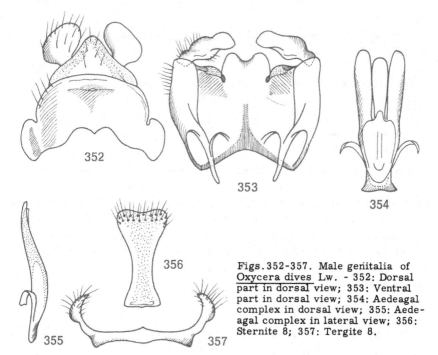

Figs. 352-357. Male genitalia of Oxycera dives Lw. - 352: Dorsal part in dorsal view; 353: Ventral part in dorsal view; 354: Aedeagal complex in dorsal view; 355: Aedeagal complex in lateral view; 356: Sternite 8; 357: Tergite 8.

and the second from Kuolajärvi (=Salla), Ks, 24.VII.1934 (leg.Krogerus). - In the mountains of C.Europe, the region of Leningrad and eastern Fennoscandia. - July-August.

33. OXYCERA FALLENII Stæger, 1844
Figs. 330, 358-361.

Oxycera fallenii Stæger, 1844: 410.

This species differs from the closely-related meigenii by having the longitudinal stripes on mesonotum not joined to the humeral spots. Margin of mesonotum above notopleural line broadly yellow from humeri to suture. Elongate and oblique lateral abdominal markings usually not narrowed at margin of abdomen.

Figs. 340-351. Morphology of Oxycera-larvae. 340: Anterior end in dorsal view of O.pardalina Meig.; 341: Abdominal segment 3 in dorsal view of same; 342: Scheme of setae on lateral wall of abdominal segment 3 of same; 343: Abdominal segment 3 in ventral view of same; 344: Posterior end in ventral view of same (float-hairs omitted); 345: Plates of posterior spiracular opening of same; 346: Posterior end in ventral view of O.formosa Meig.; 347: Plates of posterior spiracular opening of same; 348: Posterior end in ventral view of O.meigenii Stæg.; 349: Plates of posterior spiracular opening of same; 350: Posterior end in ventral view of O.pygmaea Fall.; 351: Plates of posterior spiracular opening of same. Abbreviations: see p.132.

- Male genitalia: Epandrium simple, medial process of synsternite bilobed, aedeagal complex narrowed basally. - Length: body 5.0-8.0 mm, wing: 5.8-6.2 mm.

The larval skin forming a puparium has been described by Lenz (1923). Ventral hooks present, and anal segment rounded posteriorly. Ventral spiracular plate slightly emarginate on lower margin. Penultimate segment with 3 pairs of Av setae, and 5 pairs of ventral setae distinct on anal segment. - Length: 12.0 mm, width 4.0 mm.

O. fallenii was described by Stæger (1844) as the species previously recorded by Fallén (1817) and Zetterstedt (1842) under the name of hypoleon (L.). In addition to several localities in Denmark (on E. Jutland and NE Zealand), fallenii extends into S. Sweden and rarely also into Uppland. Not found in Norway or eastern Fennoscandia at present. - C. and N. Europe, Siberia. Probably a boreal species, distributed as far south as Switzerland, Austria and Italy. - June-August. - Larvae live in clear water in streams and torrents, and adults around these habitats.

34. OXYCERA FORMOSA Meigen, 1822
Figs. 331-332, 346-347, 362-365.

Oxycera formosa Meigen, 1822: 127.

A small species with shining mesonotum and entirely yellow scutellum and legs. Longitudinal stripes on female mesonotum joined to the humeral spots. Lateral abdominal markings large, fused along the margin of abdomen. - Male genitalia: Epandrium narrowly extended laterally; medial process of synsternite simple, round. Aedeagal complex long, upcurved basally, with parameres that are forked at tips. - Length: body 3.0-4.3 mm, wing 3.0-4.0 mm.

The larva was briefly described by Lenz (1923), and its diagnostic characters were pointed out by Vaillant (1951). Yellowish, head and spiracular plates brown. Ventral hooks absent, as well as accessory Av setae. V1 and V3 setae of abdominal segments plumose but relatively short. Anal segment slightly lobate posterolaterally, with 3 distinct pairs of ventral setae. Ventral spiracular plate with a medium notch on lower margin. The number of incrustate setae on body-segments variable. - Length: 8.0 mm.

The species has been taken in several provinces of Denmark (EJ, NEZ, SZ, LFM), but apart from Denmark only 2 males have been captured in Sweden up to the present: from Nosaby and Råå, Sk. - June-July. - Larvae often in mosses growing in marshes where there is a slow water-current (Brindle, 1964b).

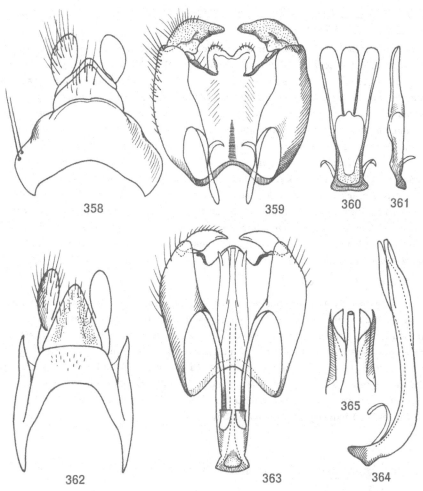

Figs. 358-365. Male genitalia of 358-361: Oxycera fallenii Stæg. and 362-365: O. formosa Meig. - 358 and 362: Dorsal part in dorsal view; 359 and 363: Ventral part in dorsal view; 360 and 365: Aedeagal complex in dorsal view; 361 and 364: Aedeagal complex in lateral view.

35. OXYCERA FREYI (Lindner, 1938)

Figs. 333-335, 366-371.

Oxycera maculata Zetterstedt, 1838: 576 (nec maculata Olivier, 1811; nec maculata Meigen, 1822).

Oxycera centralis Frey, 1911: 7 (nec centralis Loew, 1863).
Hermione freyi Lindner, 1938: 175.

A species closely related to pardalina according to general appearance and colour but differing considerably by the male genitalia. Male with thinly haired eyes, narrow longitudinal stripes on mesonotum, and scutellum black except for spines. Lateral abdominal markings almost pointed medially, triangular and broadly separated at margin of abdomen. Female with transversely-elongate upper postocular spots and conspicuous longitudinal stripes on mesonotum. Yellow notopleural stripe only very narrow, oblique pteropleural band absent, small triangular lateral spots in front of suture well-developed. Scutellum brown with a transversely-oval pale spot. In addition to rather narrow and separated lateral abdominal markings, a distinct basal spot seems to be characteristic. A median spot present on tergite 3 in the holotype of centralis, which is probably not constant. - Male genitalia: Epandrium low and wide, medial process of synsternite bilobed. Aedeagal complex rather short and simple. - Length: body 5.0-5.5 mm, wing 4.0-4.5 mm.

The larva is not known.

At present only 4 specimens are known from Sweden and Finland. A female described by Zetterstedt (1838) from Lapland and labelled "O. maculata Zett. Ins. Lapp. ♀" is the holotype of maculata (in ZIL). A further female was taken by Zetterstedt (1842) in Ly. Lpm and also in Lund. The female holotype of centralis Frey (=freyi Lindner) is labelled "Ta, Orivesi, 10. Juli 1866 (J. Sahlberg), Oxycera centralis n. sp., Frey det." and is in ZMH together with a single male that was, however, captured much later ("Ob, Pisavaara Naturpark, 12. VII. 1950, leg. H. Lindberg"). Probably a boreal species.

36. OXYCERA LEONINA (Panzer, 1798)
    Figs. 39, 372-375.

Stratiomys leonina Panzer, 1798: 12.

A species characterized by a black abdomen with only a basal and an apical yellow spot. Wings without a median cloud. Both sexes with a black mesonotum on which only the notopleural lines, post-alar calli and scutellum are yellow. - Male genitalia: Epandrium almost semicircular, bearing oval lateral plates with several long bristles. Medial process of synsternite rather deeply notched medially. Aedeagal complex of medium length, simple. - Length: body 5.5-8.0 mm, wing 5.4-6.2 mm.

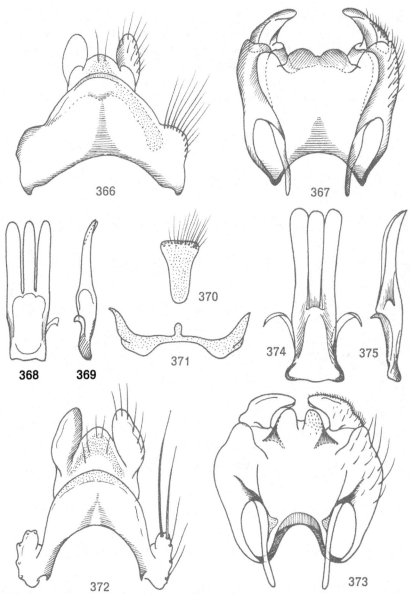

Figs. 366-375. Male genitalia of 366-371: Oxycera freyi (Lind.) and 372-375: O. leonina (Panz.). - 366 and 372: Dorsal part in dorsal view; 367 and 373: Ventral part in dorsal view; 368 and 374: Aedeagal complex in dorsal view; 369 and 375: Aedeagal complex in lateral view; 370: Sternite 8; 371: Tergite 8.

The larva has not been described.

In the nordic area, the species reaches only as far as Denmark. The material examined includes only the 6 specimens that were mentioned by Lundbeck (1907), from several localities in NEZ (Roskilde, Ringsted, Ordrup Mose). - Europe except northern parts, as rare as in Denmark. - June-August. - Adults in damp meadows, by pools, ponds and lakes.

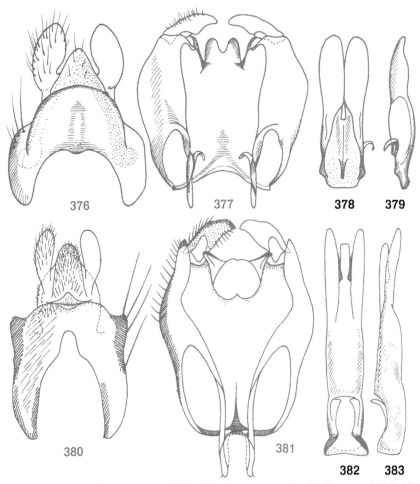

Figs. 376-383. Male genitalia of 376-379: Oxycera meigenii Stæg. and 380-383: O. pardalina Meig. - 376 and 380: Dorsal part in dorsal view; 377 and 381: Ventral part in dorsal view; 378 and 382: Aedeagal complex in dorsal view; 379 and 383: Aedeagal complex in lateral view.

37. OXYCERA MEIGENII Stæger, 1844
Figs. 336, 348-349, 376-379.

Stratiomys hypoleon L. sensu Fabricius, 1775: 760.
Oxycera meigenii Stæger, 1844: 410.

A species with conspicuously yellow pattern on head, thorax and abdomen. Yellow post-ocular band often interrupted dorsolaterally in some females like the conspicuous longitudinal stripes on thorax which may sometimes be divided into 2 pairs of yellow spots. The first pair of these is then usually joined to the yellow humeri. Lateral abdominal markings on tergites 3-4 oblique, resembling those of pulchella but usually narrower near margin of abdomen. - Male genitalia: Lateral plates of epandrium not greatly developed, medial process of synsternite rather small and bilobed. The true aedeagus apparently very reduced, and only the two large parameres distinct. - Length: body 7.0-9.0 mm, wing 6.8-8.0 mm.

The larva was described by several earlier authors (e.g. Lenz, 1923, 1926), and reliable diagnostic characters were emphasized by Schremmer (1951) and Vaillant (1951). Ventral hooks present. Posterolateral lobes of anal segment only a little prominent. 3 pairs of accessory Av setae well-developed on abdominal segments, only 2 pairs on penultimate segment. Only 3 pairs of ventral setae on anal segment. Ventral spiracular plate not emarginate. - Length: 17 mm.

Described by Stæger (1844) from Danish specimens, and stated to be the same as hypoleon sensu Fabricius. Rare but rather widely distributed in Denmark (NE Zealand, Fyn, Lolland, Falster). Apart from this, only one male taken in S. Sweden (Åkarp, Sk., 5.7.1954 - cf. Andersson, 1962). Not known from Norway and Finland. - Continental Europe, with the northern limit of its range in S. Sweden. - June-August. - Larvae in clear running water, as typical representatives of the hygropetric fauna (Vaillant, 1951), but also in saline pools and streams (Lenz, 1926). Adults on vegetation around the larval habitats.

38. OXYCERA PARDALINA Meigen, 1822
Figs. 48, 326-327, 340-345, 380-383.

Oxycera pardalina Meigen, 1822: 128.
Oxycera calceata Loew, 1871: 41.

A very variable species. Male always with unstriped mesonotum, scutellum all yellow, legs mainly yellow and abdomen with yellow lateral markings united at the margin of abdomen. Some specimens (including a single male taken in Sweden) have scutellum almost black with yellow spines only, legs predomi-

113

nantly brown, and orange abdominal lateral markings only indistinctly joined at margin. In females there is variation in the shape of longitudinal frontal stripes (short, long or divided and then with a distinct upper pair of rounded spots); hind legs may be darkened; and abdominal lateral markings may be narrow, fused laterally, or strikingly large and even joined medially on tergite 4, but in other cases they may be small and well-separated. - Male genitalia: Epandrium with somewhat produced anterior angles, bearing several long bristles. Medial process of synsternite low, with distinct lateral corners. Aedeagal complex long but straight. - Length: body 4.5-5.0 mm, wing 4.2-5.6 mm.

The larva was described by Lenz (1923, 1926) and Schremmer (1951) under the name of calceata, and the important diagnostic features have also been pointed out by Bertrand (1948) and Vaillant (1951). Ventral hooks on penultimate segment relatively strong. Posterolateral lobes of anal segment well developed and pointed apically. 3 pairs of accessory Av setae present on abdominal segments 1-6 and 1 pair on segment 7. V1 and V2 setae of anal segment absent. Lower margin of the ventral spiracular plate deeply emarginate. Numerous incrustate setae on body segments. - Length: 10.0-11.0 mm, maximum width: 2.5-3.0 mm (many larvae and puparia from Central Europe).

In Scandinavia, a pair of specimens was first taken by Zetterstedt (1842) in Småland, Sweden. A further 3 females, found in Kinnekulle, Vg., were recorded under the name of pulchella by Wahlgren (1907). No further material is known from the nordic area. All over Europe, but rare in the North. - June-July. - The hygropetric larvae live among algae and in wet moss cushions, preferring the clear water of springs and streams. Adults around the larval habitat, on vegetation, flowers, stones and rocks.

39. OXYCERA PYGMAEA (Fallén, 1817)
Figs. 337-338, 350-351, 384-387.

Stratiomys pygmaea Fallén, 1817: 11.

A relatively tiny species, resembling formosa in particular, but without longitudinal stripes on mesonotum in either sex and with dull black thorax in male. Head of female mainly yellow, but rarely partly darkened. - Male genitalia: Posterolateral angles of epandrium slightly projecting, usually with only 2 bristles. Medial process of synsternite simply rounded. Dististyles sharply pointed. Aedeagal complex short and wide, with rather stout basal part. - Length: body 3.0-4.2 mm, wing 3.5-4.0 mm.

The larva was described by Vaillant (1951). Ventral hooks conspicuous and long. Anal segment with prominent and pointed posterolateral lobes. 1-2 pairs of accessory Av setae on abdominal segments 2-5 only. Ventral spiracular plate distinctly notched on lower margin but not as deeply as in pardalina. - Length: 7.0 mm.

Rare in northern Europe, and found only in Denmark and Sweden (in addition to England). In Sweden, described by Fallén (1817) from Östergötland, and confirmed in Sk., Sm., Gtl., and Jmt. by further material. In Denmark found at several localities in NE and S Zealand, and Bornholm. - June-August. - Mainly C. and N. Europe, from the Pyrenees and the Alps to England and C. Sweden. - Larvae in clear running water. Adults around the larval habitats.

40. OXYCERA TERMINATA Meigen, 1822
Figs. 388-391.

Oxycera terminata Meigen, 1822: 130.

This species is characterized by having only an apical yellow spot on a black abdomen, and can be confused only with leonina. Antennae brown, basal segments usually yellow, and legs almost wholly yellow. Female head with lateral post-ocular spots in addition to the usual upper pair of yellow patches. - Male genitalia: Epandrium almost rectangular, with a deep basal notch, medial process of synsternite rather high and simple. Aedeagal complex lyre-shaped in apical part. - Length: body 4.0-6.0 mm, wing 4.2-5.2 mm.

The larva is not known.

A very rare species, taken only on Bornholm in Denmark and apparently not occurring in Fennoscandia. Only one of the 3 female specimens taken at Hasle, 10. VI. 1870 by Schlick still preserved in ZMC, cf. also Lundbeck, 1907). - Throughout most of Europe, from the Pyrenees, Jugoslavia and Greece to Great Britain, Denmark and the Leningrad Region. - Adults found along streams.

41. OXYCERA TRILINEATA (Linné, 1767)
Figs. 325, 392-395.

Musca trilineata Linné, 1767: 980.
Musca hypoleon Linné, 1767: 980.

Unlike the other nordic species of Oxycera, trilineata has the ground-colour green or yellow with black pattern. This pattern is rather variable, on the ab-

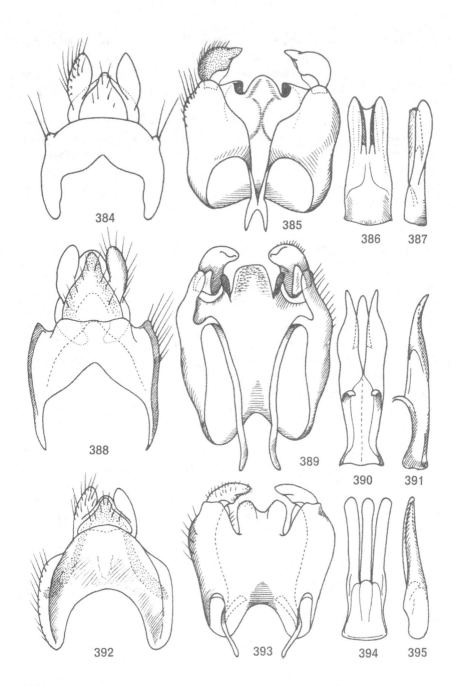

384

385

386 387

388

389

390 391

392

393

394 395

domen in particular is sometimes reduced to relatively narrow transverse stripes or is fused longitudinally to a varying extent. - Male genitalia: Epandrium almost semicircular, medial process bilobed with rounded middle notch, aedeagal complex simple, parameres well separated. - Length: body 5.0-7.0 mm, wing 4.2-6.0 mm.

The larva was described by Heeger (1856) and Lenz (1926), and was also mentioned by Lundbeck (1907) and Brindle (1964b). Apparently without ventral hooks and with the anal segment rounded posteriorly. Abdominal segments with plumose, branched setae V1, V3 and Vl. Ventral spiracular plate not emarginate on lower margin. - Length: 9.0-10.0 mm, maximum width: 2.0-3.0 mm.

The commonest species of the genus. Known from several provinces in Denmark and Sweden (N. to Ög. and Boh.) but rarely taken in Norway (Oslo, Tøien, Ak) and Finland (Aland Islands only); also in the Soviet Ib. - All over the Palaearctic Region, from North Africa and Europe to China. - The larva was found at the edge of a torrent (Heeger, 1856) and in mud (Lundbeck, 1907), in marshes near a stream and in a ditch with saline water (Lenz, 1926). Adults on ground-vegetation near water.

Subfamily Pachygasterinae

Mostly tiny and mainly black species. Third antennal segment round or oval, consisting of 4-5 flagellomeres and with a terminal or subterminal 2- to 3-segmented arista. Posterior margin of scutellum without spines. Only 3 M-veins arising from the large discal cell, and vein R4+5 seldom unbranched. Abdomen short and broad, consisting of 5 main segments. Male genitalia highly specialized. Cerci often with groups of long setae, sometimes bilobed. Synsternite compact, medial process of synsternite virtually absent. Aedeagal complex slender and long.

The larvae have the characteristic shape of terrestrial forms. Last abdominal segment rounded posteriorly and anal slit bordered by fine lateral teeth in some genera. A mid-ventral patch on segment 6 always distinct, round or elongate-oval. Vl setae located rather ventrally, sometimes 5 setae on lateral wall

Figs. 384-395. Male genitalia of 384-387: Oxycera pygmaea (Fall.), 388-391: O. terminata Meig. and 392-395: O. trilineata (L.). - 384, 388 and 392: Dorsal part in dorsal view; 385, 389 and 393: Ventral part in dorsal view; 386, 390 and 394: Aedeagal complex in dorsal view; 387, 391 and 395: Aedeagal complex in lateral view.

of abdominal segments. The larvae are usually found under the bark of dead trees, in tree-holes, etc.

Six European species, the differences between which are considered as being of generic rank following the treatment of the Nearctic representatives of the subfamily (cf. Kraft & Cook, 1961; James, 1965).

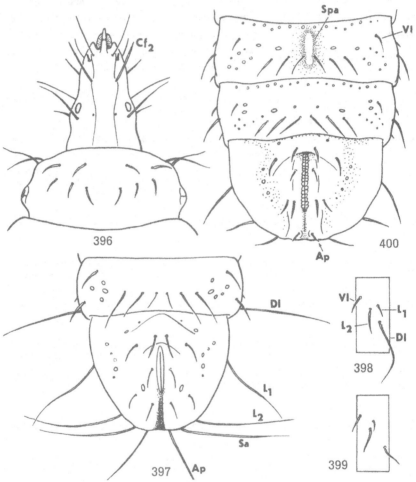

Figs. 396-400. Morphology of larvae of Pachygastrinae. - 396: Anterior end in dorsal view of Zabrachia minutissima (Zett.); 397: Posterior end in ventral view of same; 398: Scheme of setae on lateral wall of abdominal segment 3 of same; 399: Same of Neopachygaster meromelaena (Duf.); 400: Posterior end in ventral view of same. Abbreviations: see p. 132.

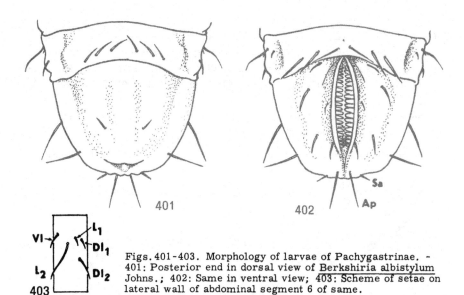

Figs. 401-403. Morphology of larvae of Pachygastrinae. -
401: Posterior end in dorsal view of Berkshiria albistylum
Johns.; 402: Same in ventral view; 403: Scheme of setae on
lateral wall of abdominal segment 6 of same.

Genus Pachygaster Meigen, 1803

Pachygaster Meigen, 1803, Mag.Insektenkd., 2: 266.
Vappo Latreille, 1804, Tabl.meth.Ins., 24: 193.
Alphapachygaster Pleske, 1922, Ann.Mus.Zool.Ac.Sci.Russ., 23: 336.
Type-species: Nemotelus ater Panzer, 1798.

Post-ocular part of head wide in both sexes. Scutellum triangular to semicircular, without any marginal rim. Cross-vein r-m present.

Nearctic larvae of the genus have no prominent teeth along anal slit and lateral setae of anal segment long, at least one-third as long as width of anal segment. Thoracic leg-groups consisting of 3 setae; only 3 setae on lateral wall
of abdominal segments.

Only one Palaearctic species:

42. PACHYGASTER ATRA (Panzer, 1798)
    Figs. 404-411.

Nemotelus ater Panzer, 1798: 5.
Sargus pachygaster Fallén, 1817: 13.

In addition to the characters mentioned above, the following features are of diagnostic importance: Antennae black in male and yellow in female, basal half

119

of wing distinctly darkened, knob of halteres blackish, femora black. - Male genitalia: Cerci with basal tuft of long setae, synsternite notched posteromedially, dististyles pointed at tips, with a small inner lobe. Aedeagal complex slender and rather simple in apical part. - Length: body 3.0-4.0 mm, wing 2.8-3.8 mm.

The larva was superficially described by Heeger (1853) and its general appearance was illustrated by Johannsen (1922).

The species has only been recorded in a few Swedish localities (Sk., Gtl.,

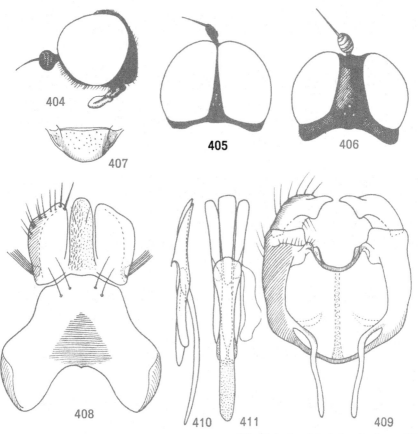

Figs. 404-411. Pachygaster atra (Panz.). - 404: Male head in lateral view; 405: Male head in dorsal view; 406: Female head in dorsal view; 407: Male scutellum; 408: Dorsal part of male genitalia in dorsal view; 409: Ventral part of male genitalia in dorsal view; 410: Aedeagal complex in lateral view; 411: Aedeagal complex in dorsal view.

Ög.), and no further records are known from Denmark and Fennoscandia. - Throughout most of Europe but obviously rare in the North. - June-August. - The larvae are known to occur in various substrates, e.g. under the bark of rotten trees (Populus alba, Salix alba, Ulmus), in decaying leaves beneath a walnut-tree, and in leaf-mould. The mass-occurrence of adults is not uncommon in woods, on leaves of bushes, etc.

Note: Some previous authors (e.g. Bezzi, 1903; Verrall, 1909) have thought that Fallén's S.pachygaster is in fact a mixture of atra and tarsalis, but all 9 specimens labelled by Fallén "Sarg.pachygaster m.Fall." in the Stockholm Museum belong without any doubt solely to atra.

Genus Praomyia Kertész, 1916

Praomyia Kertész, 1916, Annls Mus.nat.hung., 14: 184.
Type-species: Pachygaster leachii Curtis, 1824.

Post-ocular part of head little prominent on basal half. Upper eye-facets in male larger than the lower. Scutellum semicircular, with hardly any rim. Cross-vein r-m distinct.

Larva with short Dl setae on abdominal segments. Anal segment with relatively long Sa setae and shorter but clearly visible Ap setae.

Probably 3 species in the Palaearctic Region, but only one in C.and N.Europe.

43. PRAOMYIA LEACHII (Curtis, 1824)
    Figs.412-419.

Pachygaster leachii Curtis, 1824: 42.
Pachygaster pallipennis Macquart, 1834: 265.

A species with yellow to orange antennae, the basal segments sometimes darkened. Wings entirely clear, only anterior veins sometimes brownish. Knob of halteres usually blackish in male, but yellowish at least at apex in female. Legs predominantly yellow, usually with a black preapical ring on hind femora. - Male genitalia: Cerci bilobed, with a basal tuft of long setae as in atra. Synsternite not notched posteriorly, dististyles rather stout, with blunt tips. Aedeagal complex of almost the same shape as in atra. - Length: body 3.0-3.5 mm, wing 2.8-3.5 mm.

The larva was figured by Séguy (1926) and Brindle (1962). The latter author pointed out the main diagnostic characters.

121

So far only known from Sweden, viz. in Gotland (Zetterstedt, 1842) and Öland (Wahlberg, 1854); its occurrence in Denmark and S. Finland seems quite possible. Frey (1941) listed the species from Finland but this record is very probably based on a note by Pleske (1924) concerning the locality Moustanijaki, allegedly situated in S. Finland but nowadays in the USSR (Moustanijaki=Mustam-

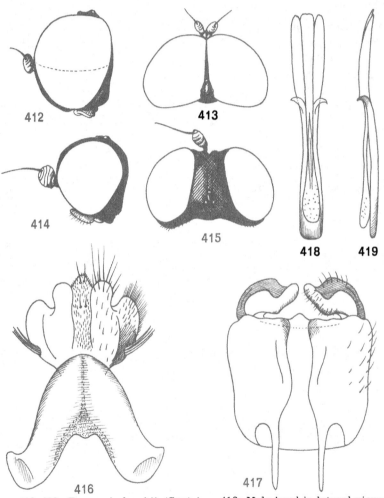

Figs. 412-419. Praomyia leachii (Curt.). - 412: Male head in lateral view; 413: Male head in dorsal view; 414: Female head in lateral view; 415: Female head in dorsal view; 416: Dorsal part of male genitalia in dorsal view; 417: Ventral part of male genitalia in dorsal view; 418: Aedeagal complex in dorsal view; 419: Aedeagal complex in lateral view.

jaki=Gorkovskoje, cf. Stackelberg, 1954). - Throughout most of Europe and Transcaucasia. - June-August. - Larvae under bark of oaks; Smith (1957) found larvae at the roots of Angelica in England. Adults mostly on leaves of shrubs and trees in sunny places.

Genus Neopachygaster Austen, 1901

Neopachygaster Austen, 1901, Entomologist's mon. Mag., 12: 245.
Type-species: Pachygaster meromelas Dufour, 1841.

Antennae inserted above middle of the head-profile. Third antennal segment with a large shining area on inner surface. Eyes separated in both sexes. Depressed part of lower frons, above each antennal base, lined with dense silver pile. Post-ocular part of head extremely narrow. Scutellum rather small and semicircular. Cross-vein r-m distinct.

Larvae with scarcely distinct anal teeth and relatively short setae. Ap setae of anal segment in particular remarkably short, hardly visible in dorsal view.

Only one Palaearctic species:

44. NEOPACHYGASTER MEROMELAENA (Dufour, 1841)
   Figs. 399-400, 420-428.

Pachygaster meromelas Dufour, 1841: 266.
Pachygaster orbitalis Wahlberg, 1854: 212.

Head very high in both sexes, antennae orange, darkened apically; silvery-white bands along eye-margin on face continued as round spots above antennae in both sexes. Wings clear, knob of halteres yellowish. Femora black, and tibiae sometimes also darkened in middle. - Male genitalia: Epandrium remarkably long, fused with epiproct and cerci. Cerci with 2-3 more or less distinct tufts of setae. Synsternite also rather long. Dorsal bridge produced into a cylindrical hyaline cover for aedeagal complex. - Length: body 2.5-3.0 mm, wing 2.5-3.2 mm.

The larva was noted by Dufour (1841) and subsequently by several other authors. A detailed description was published by Krivosheina (1965). Sa setae on anal segment shorter than L setae, Ap setae extremely short. - Length: 4.5-5.8 mm, maximum width: 0.8-1.3 mm. (Extensive material from Central Europe).

Apparently not common in Fennoscandia. Described as <u>orbitalis</u> by Wahlberg
(1854) from Sweden, Gusum, Ög., and later also found on Gotska Sandön. Its
occurrence in Finland is verified by material from Föglö, Bänö, Al; Helsinge,
N. - Probably throughout most of Europe, but its range is inadequately known

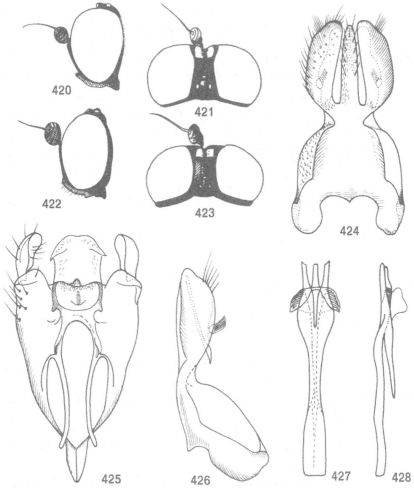

Figs. 420-428. Neopachygaster meromelaena (Duf.). - 420: Male head in lateral view; 421: Male head in dorsal view; 422: Female head in lateral view; 423:
Female head in dorsal view; 424: Dorsal part of male genitalia in dorsal view;
425: Ventral part of male genitalia in dorsal view; 426: Dorsal part of male genitalia in lateral view; 427: Aedeagal complex in dorsal view; 428: Aedeagal
complex in lateral view.

(W. Europe, Fennoscandia, Czechoslovakia, NW USSR). - June-July. - The host-plant specifity of the larva is apparently not as strong as Verrall (1909) believed and in addition to holly (Ilex), from which the species has been bred in England, it has also been found under the decaying bark of other trees (Populus alba, P. tremula).

Genus Eupachygaster Kertész, 1911

Eupachygaster Kertész, 1911, Mém. Ier Congr. intern. ent., 2: 31.
Type-species: Pachygaster tarsalis Zetterstedt, 1842.

Post-ocular part of head in dorsal view virtually invisible in male and only slightly projecting in female. Upper eye-facets in male much larger than the lower. Scutellum as long as wide, paraboloid and with a well-developed rim, with many marginal spinules. Cross-vein r-m practically absent.

Larva with long L2 setae and equally long L, Sa and Ap setae on anal segment. Usually 4 setae on lateral wall of abdominal segments. D3 on abdominal segments conspicuously short, only 1/4 length of segment.

Only one Palaearctic species:

45. EUPACHYGASTER TARSALIS (Zetterstedt, 1842)
   Figs. 41, 429-438.

Pachygaster tarsalis Zetterstedt, 1842: 152.

Basal antennal segments always yellow, third segment often darkened, especially in females. Two silvery-grey spots above antennae in female. Basal half of wings contrasting blackish, knob of halteres also black. Femora black, tibiae and tarsi pale yellow. - Male genitalia: Epandrium high and fused with epiproct, cerci with an apical tuft of setae. Synsternite compact and with a large dorsal bridge forming a low hyaline cover for aedeagal complex. Dististyles tapered towards tips with a low basal lobe. Aedeagus strongly tapering in basal part. - Length: body 3.0-4.5 mm, wing 2.8-4.3 mm.

Larva: The general appearance of a puparium was illustrated by Verrall (1909) and Séguy (1926). The main diagnostic characters, in comparison with some Nearctic species, are mentioned above.

Zetterstedt (1842) described the species from material taken at Rohne and Westerby near Thorsborg (9. and 13. VII. 1841). Danish specimens were bred

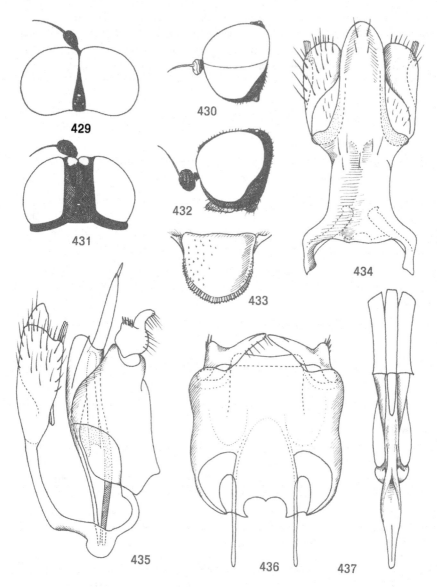

Figs. 429-437. Eupachygaster tarsalis (Zett.). - 429: Male head in dorsal view; 430: Male head in lateral view; 431: Female head in dorsal view; 432: Female head in lateral view; 433: Male scutellum; 434: Dorsal part of male genitalia in dorsal view; 435: Male genitalia in lateral view; 436: Ventral part of male genitalia in dorsal view; 437: Aedeagal complex in dorsal view.

from larvae found at several localities (Charlottenlund, Dyrehaven and Maribo). For the present, no further material is known from the nordic area. - C. and N. Europe, reaching as far south as Jugoslavia and Bulgaria, and N to S. Sweden. - June-July. - Larvae have been found under bark and in the decaying wood of beeches, oaks and apple-trees.

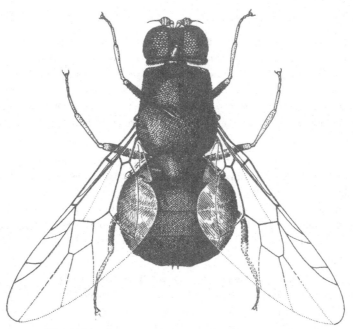

Fig. 438. Female of Eupachygaster tarsalis (Zett.).

Genus Zabrachia Coquillett, 1901

Zabrachia Coquillett, 1901, Bull. N. Y. St. Mus. Sci. Serv., 47: 585.
Type-species: Zabrachia polita Coquillett, 1901.

Head almost globular, the posterior part hardly visible at posteroventral angles in male and slightly projecting in female. Scutellum small and triangular. Vein R4+5 simple, unbranched.

Larvae with very long L2 setae on abdominal segments, like all the marginal setae on anal segment. The dorsal setae are equal in length, also D3 are as long as the other dorsal setae.

Only one species in the Palaearctic Region:

46. ZABRACHIA MINUTISSIMA (Zetterstedt, 1838)
Figs. 40, 49, 396-398, 439-447.

Pachygaster minutissima Zetterstedt, 1838: 575.

A small shining black species. The hyaline wings, with unbranched R4+5 are the most striking characters. Antennae black in male and orange in female, female eyes widely separated. Knob of halteres white. - Male genitalia: Dorsal part simple, cerci small, without distinct groups of longer setae. Synsternite with straight posterior margin, dististyles simple, leaf-shaped. Aedeagal complex long and slender, parameres forked. - Length: body 2.0-3.0 mm, wing 1.8 -2.8 mm.

The larva was rather superficially described by several earlier authors, and some figures are included in Hennig's work (1952). Detailed descriptions have been published by Brindle (1962) and Krivosheina (1965). Yellowish to brown, with a darker pattern consisting of spots arranged in two curved lines along the edges. The main diagnostic characters are mentioned above. - Length: 5.0-6.0 mm, maximum width: 0.9-1.5 mm. (2 puparia in ZMH and extensive material from Central Europe).

The species was described by Zetterstedt (1838) from C. Sweden (Dlr., leg. Boheman) and was later taken in other Swedish provinces (Sm., G. Sand., Ög., Upl., Jmt.). It is known to occur in several provinces of eastern Fennoscandia, including the Soviet part. Danish specimens were mostly bred from larvae found on Zealand, but one male was captured at Engestofte (LFM). - C. and N. Europe including N. Fennoscandia. - June-August. - Larvae live under the bark of coniferous trees (Picea, Abies, Pinus), and often under the bark of spruce stumps.

Genus Berkshiria Johnson, 1914

Berkshiria Johnson, 1914, Psyche, 21: 158.
Pseudowallacea Kertész, 1921, Annls Mus.nat.hung., 18: 168, syn.n.
Type-species: Berkshiria albistylum Johnson, 1914.

The type-species of the two monotypic genera Berkshiria and Pseudowallacea are identical, as has been proved by recent examination of the external mor-

phology, male genitalia and larval characters of European and American specimens. The monotypic genus <u>Berkshiria</u> is therefore Holarctic in distribution.

Head conspicuously higher than wide in lateral view. Eyes holoptic in male
and dichoptic in female. Third antennal segment consisting of 5 main flagellomeres in the form of an elongate-oval flagellum and 3 slender apical flagello-

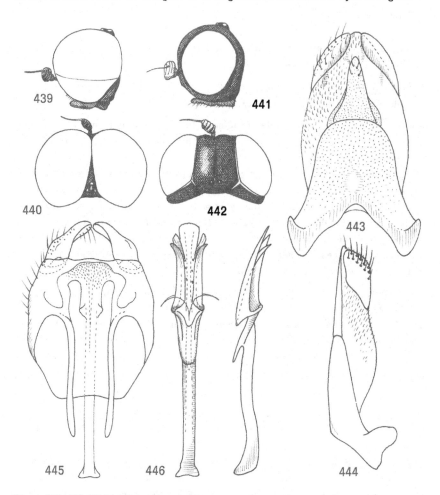

Figs. 439-447. Zabrachia minutissima (Zett.). - 439: Male head in lateral view;
44o: Male head in dorsal view; 441: Female head in lateral view; 442: Female
head in dorsal view; 443: Dorsal part of male genitalia in dorsal view; 444:
Same in lateral view; 445: Ventral part of male genitalia in dorsal view; 446:
Aedeagal complex in dorsal view; 447: Aedeagal complex in lateral view.

meres forming a terminal, whitish pubescent arista. Scutellum distinctly marginate, with a wide posterior rim. R4 usually present, but partially or completely absent in some specimens.

Larvae with conspicuous anal teeth and relatively short setae on body segments. Sternal patch on mid-ventral line of abdominal segment 6 always round.

Only one species:

47. BERKSHIRIA ALBISTYLUM Johnson, 1914
Figs. 401-403, 448-456.

Berkshiria albistylum Johnson, 1914: 158.
Pseudowallacea hungarica Kertész, 1921: 170, syn.n.

Antennae inserted above middle of the head-profile, mainly brown with darkened basal segments and black tips, arista whitish pubescent. The narrow silvery-white stripes along eye-margin in female reaching just above antennae. Wings quite clear, R4 present but often incomplete. Halteres and squamae black. Legs black, with metatarsus and usually also second segment yellowish to white. - Male genitalia: Short pointed surstyles at posterolateral angles of epandrium usually rather distinct. Cerci simple, without special groups of longer setae. Synsternite elongate, dorsal bridge developed. Aedeagal complex long, slender and flat, parameres covered with spinules on lateral and ventral sides. - Length: body 3.5-4.0 mm, wing 3.5-3.8 mm.

The larva was adequately described by Kraft & Cook (1961) and Krivosheina (1965). Anal slit surrounded by a row of long teeth. Abdominal segments 1-6 usually with 5 setae on lateral wall, but accessory Dl setae very small or even absent. Anal segment with moderately long L1, L2 and Ap setae, but relatively short Sa setae. - Length: 5.0-7.4 mm, maximum width: 1.8-2.0 mm. (Several puparia without anterior part in NRS).

European specimens have been determined as Pseudowallacea hungarica Kert. in the past. The original description of this species was based on two females from Roumania. The male was described later by Hanson (1942) from material taken in Sweden (Vassunda, Tursbo, Upl.; Strömsnäs, Nrk.). There is also one female in NMG (Aneboda, Sm.), while several specimens from the localities mentioned above are in NRS. In Finland the species was listed by Frey (1941) and recorded from Karislojo, Ab and Esbo, N; it was also found in Soviet Carelia (Impilahti, Kr). - In Europe known only from Roumania, Fennoscandia and NW USSR. Widely distributed in eastern North America, from northern Alabama to southern Ontario. - May-June. - Larvae have been collected from under the bark of aspen trunks (<u>Populus tremula</u>) in Europe and under the bark of <u>P</u>. <u>deltoides</u> and <u>Ulmus pumila</u> in North America.

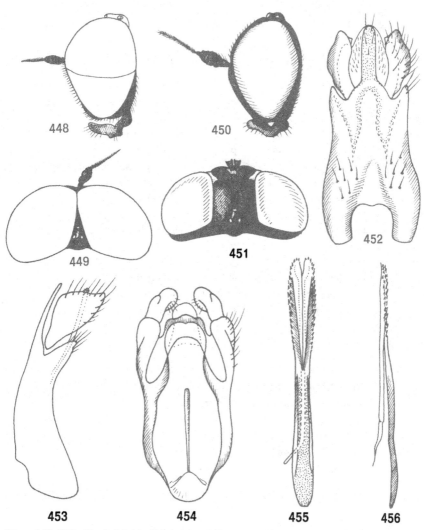

Figs. 448-456. Berkshiria albistylum Johns. - 448: Male head in lateral view; 449: Male head in dorsal view; 450: Female head in lateral view; 451: Female head in dorsal view; 452: Dorsal part of male genitalia in dorsal view; 453: Same in lateral view; 454: Ventral part of male genitalia in dorsal view; 455: Aedeagal complex in dorsal view; 456: Aedeagal complex in lateral view.

Abbreviations

Morphological terms:

A	- antennae		Lb	- labral setae
Ad	- anterodorsal setae		Lp	- lateral cuticular plates
ae	- aedeagus		Mm	- mandibular-maxillary complex
Ap	- apical setae		mp	- medial process
As	- anterior spiracle		p	- paramere
Asl	- anal slit		pp	- paraproct
Av	- anteroventral setae		pr	- proctiger
c	- cerci		Ps	- posterior spiracular opening
Cf	- clypeofrontal setae		Psp	- pupal spiracles
D	- dorsal setae		s	- synsternite
Dl	- dorsolateral setae		Sa	- subapical setae
dp	- dorsal process		Sp	- rudimentary spiracles
dst	- dististyle		Spa	- sternal patch
Dt	- dorsal tubercles		ss	- surstyli
e	- epandrium		vl	- ventral lobe
Ep	- eye prominence		Vl	- ventrolateral setae
L	- lateral setae		V	- ventral setae

Museums:

NMG - Natural History Museum, Göteborg, Sweden.
NRS - Natural History Museum, Stockholm, Sweden.
ZIL - Zoological Institute, Lund, Sweden.
ZMB - Zoological Museum, Bergen, Norway.
ZMC - Zoological Museum, Copenhagen, Denmark.
ZMH - Zoological Museum, Helsinki, Finland.
ZMO - Zoological Museum, Oslo, Norway.

Literature

Andersson, H., 1962: Sällsynta eller för Sverige nya tvåvingar (Diptera). - Opusc.ent., 27: 62-64.

- 1971a: Faunistic notes on Scandinavian Diptera Brachycera. - Ent.Tidskr., 92: 232-236.

- 1971b: Taxonomic notes on Scandinavian species of the genus Sargus Fabr. (Dipt., Stratiomyidae). - Ent.scand., 2: 237-240.

Ardö, P., 1957: Studies in the marine shore dune ecosystem with special reference to the dipterous fauna. - Opusc.ent., Suppl.14: 255 pp., Lund.

Austen, E.E., 1899: On the preliminary stages and mode of escape of the imago in the dipterous genus Xylomyia, Rond. (Subula, Mg.et auct.), with especial reference to Xylomyia maculata, F.; and on the systematic position of the genus. - Ann.Mag.nat.Hist., ser.7, 3: 181-190.

Bertrand, H., 1948: Notes sur deux larves du genre Hermione Meigen. - Bull. Soc.ent.Fr., 53: 55-58.

Bezzi, M., 1903: Orthorrhapha Brachycera, in: Katalog der paläarktischen Dipteren. Vol.II. Budapest.

Bidenkap, O., 1900: Foreløbig Oversigt over de i det arktiske Norge hidtil bemærkede Diptera Brachycera. - Tromsø Mus.Årsh., 23: 1-112.

Bischoff, W., 1925: Über die Kopfbildung der Dipterenlarven. III. Teil. Die Köpfe der Orthorrhapha-Brachycera-Larven. - Arch.Naturgesch., 90, A 8: 49-105.

Bonsdorff, E.J., 1861: Finlands tvåvingade Insekter (Diptera), förtecknade och i korthet beskrifne. - I-XII + 37-301 pp. Helsingfors.

Bouché, P.F., 1834: Naturgeschichte der Insekten, besonders in Hinsicht ihrer ersten Zustände als Larven und Puppen. - 216 pp., Berlin.

Brindle, A., 1961: Taxonomic notes on the larvae of British Diptera - 3. The genus Solva, Walker (Xylomyia, Rondani: Subula, Meigen) (Stratiomyidae). - Entomologist, 94: 202-205.

- 1962: Taxonomic notes on the larvae of British Diptera - 7. The genus Pachygaster Meigen (Stratiomyidae). - Entomologist, 95: 77-82.

- 1964a: Taxonomic notes on the larvae of British Diptera - 16. The Stratiomyinae (Stratiomyidae). - Entomologist, 97: 92-96.

- 1964b: Taxonomic notes on the larvae of British Diptera - 17. The Clitellariinae (Stratiomyidae). - Entomologist, 97: 134-139.

- 1965: Taxonomic notes on the larvae of British Diptera - 23. The Geosarginae (Stratiomyidae). - Entomologist, 98: 208-216.

Cornelius, C., 1860: Zur Ernährung und Entwicklung der Larven von Sargus formosus Schrank. - Stettin.ent.Ztg., 21: 202-204.

Curtis, J., 1824: British Entomology: Being illustrations and descriptions of the genera of insects found in Great Britain and Ireland. Vol.1, pls. 1-50, London.

Dale, J.C., 1842: Descriptions, & c. of a few rare or undescribed species of

British Diptera, principally from the collection of J.C.Dale, Esq., M.A., F.L.S., & c. - Ann.Mag.nat.Hist., 8: 430-433.

Dufour, L., 1841: Note sur la larve du Pachygaster meromelas. - Annls Sci. nat., 16: 264-266.

- 1847: Histoire des metamorphoses du Subula citripes et quelques autres especes de ce genre des diptères. - Annls Sci.nat., 7: 5-14.

Dušek, J., 1961: Beitrag zur Kenntnis der Larven der Gattung Eulalia Meig. (Dipt.: Stratiomyiidae). - Zool.Listy, 10: 211-217.

Dušek, J. & Rozkošný, R., 1963-1967: Revision mitteleuropäischer Arten der Familie Stratiomyidae (Diptera) mit besonderer Berücksichtigung der Fauna der CSSR. I.-IV. - Acta ent.bohemoslov., 60 (1963): 201-221; 61 (1964): 360-373; 62 (1965): 24-60; 64 (1967): 140-165.

- 1966: Ergebnisse der Albanien-Expedition 1961 des Deutschen Entomologischen Institutes. - 57. Beitrag - Diptera: Stratiomyidae. - Beitr.Ent., 16: 507-521.

- 1968: Beris strobli nom.nov. (Diptera, Stratiomyidae). - Reichenbachia, 10 (39): 293-298.

Fabricius, J.C., 1775: Systema entomologiae. 832 pp., Flensburgi et Lipsiae.

- 1781: Species insectorum. Vol.2, 517 pp. Hamburgi et Kilonii.

- 1787: Mantissa insectorum. Vol.2, 382 pp., Hafniae.

- 1794: Entomologia systematica emendata et aucta. Vol.4, 472 pp. Hafniae.

- 1805: Systema antliatorum. 373 pp. + 30 pp., Brunsvigae.

Fallén, C.F., 1817: Stratiomydae Sveciae. 14 pp., Lundae.

- 1826: Supplementum Dipterorum Sveciae. 16 pp., Lundae.

Forster, J.R., 1771: Novae species insectorum. Centuria I. 100 pp., London.

Frey, R., 1911: Zur Kenntnis der Dipterenfauna Finlands. Stratiomyidae, etc. - Acta Soc.Fauna Flora fenn., 34 (6): 1-59.

- 1918: Beitrag zur Kenntnis der Dipterenfauna des nördl.europäischen Russlands. - II. Dipteren aus Archangelsk. - Acta Soc.Fauna Flora fenn., 46 (2): 1-32.

- 1941: Diptera-Brachycera (excl.Muscidae, Tachinidae), in: Enumeratio Insectorum Fenniae. - VI. Diptera. 63 pp., Helsingfors.

- 1947: Anteckningar om dipterfaunan på Karlö (Hailuoto) sommaren 1947. - Mem.Soc.Fauna Flora fenn., 24: 69-80.

- 1953: Über Oxycera Freyi Lind. (Dipt. Stratiomyidae). - Notul.ent., 33: 71-72.

- 1960: Die paläarktischen und südostasiatischen Solviden (Diptera). - Commentat.biol., 23 (1): 1-15.

Goidanich, A., 1939: Contributi alla conscenza dell'entomofauna di risaia I.- Gli Straziomiidi: mancati nemici del Riso. - Risicoltura, 29: 221-23o.

Haliday, A.H., 1857: On some remaining blanks in the natural history of the native Diptera. - Nat.Hist.Rev., 3: 177-196.

Hammer, O., 1941: Biological and ecological investigations on flies associated with pasturing cattle and their excrement. - Vidensk.Meddr dansk naturh. Foren., 105: 141-393.

Hanson, B.H., 1942: Das bisher unbekannte Männchen einer für Schweden neuen Fliegenart (Dipt. Strat.). - Ent. Tidskr., 63: 63-64.

Heeger, E., 1853: Beiträge zur Naturgeschichte der Insecten. - Sber. Akad. Wiss. Wien, 10 (2): 161-178.

- 1856: Neue Metamorphosen einiger Dipteren. - Sber. Akad. Wiss. Wien, 20: 335-350.

- 1858: Neue Metamorphosen einiger Dipteren. - Sber. Akad. Wiss. Wien, 31: 295-308.

Hennig, W., 1952: Die Larvenformen der Dipteren. Vol. 3, 628 pp., Berlin.

- 1967: Die sogenannten "niederen Brachycera" im Baltischen Bernstein. - Stuttg. Beitr. Naturk., 174: 1-51.

James, M. T., 1965: Stratiomyidae, in: Stone, A. et al., A catalog of the Diptera of America North of Mexico. 1696 pp., Washington, D.C.

Jansson, A., 1922: Faunistiska och biologiska studier över insektlivet vid Hornsjön på norra Öland. - Ark. Zool., 14 (23): 1-81.

- 1925: Die Insekten-, Myriapoden- und Isopodenfauna der Gotska Sandön. 182 pp., Örebro.

- 1935: Supplement till Die Insekten-, Myriapoden- und Isopodenfauna der Gotska Sandön. - Ent. Tidskr., 56: 52-87.

Johannsen, O. A., 1922: Stratiomyiid larvae and puparia of the northeastern states. - Jl N.Y. ent. Soc., 30: 141-153.

Johnson, C. W., 1914: A new stratiomyid. - Psyche, 21: 158-159.

Jørgensen, L., 1917: Fortegnelse over Lolland-Falsters Vaabenfluer (Stratiomyiidae). - Flora og fauna, 1917: 13.

Kertész, K., 1921: Vorarbeiten zu einer Monographie der Notacanthen. - Annls Mus. nat. hung., 18 (1920-21): 153-176.

Kraft, K. J. & Cook, E. F., 1961: A revision of the Pachygasterinae (Diptera, Stratiomyidae) of America North of Mexico. - Misc. Publs ent. Soc. Am., 3: 1-24.

Krivosheina, N. P., 1965: New data on the taxonomy of dendrophilous chameleon flies (Diptera, Stratiomyidae) and of their larvae. - Ént. Obozr., 44: 652-664. (In Russian).

Lenz, F., 1923: Stratiomyiiden aus Quellen. Ein Beitrag zur Metamorphose der Stratiomyiiden. - Arch. Naturgesch., 89, A 2: 39-62.

- 1926: Stratiomyidenlarven aus dem Salzwasser. - Mitt. geogr. Ges. naturh. Mus. Lübeck, 31: 170-174.

Lindner, E., 1936-38: Stratiomyiidae in: Lindner E., Die Fliegen der palaearktischen Region. Pt. 18, 218 pp., Stuttgart.

Lindroth, C.H., 1943: Oodes gracilis Villa. Eine termophile Carabide Swedens. - Notul. ent., 22: 109-157.

Linné, C., 1758: Systema naturae per regna tria naturae. Ed. 10., Vol. 1, 824 pp., Holmiae.

- 1761: Fauna svecica sistens animalia Sveciae regni. Ed. 2, 578 pp., Stockholmiae.

- 1767: Systema naturae per regna tria naturae. Ed. 12 (rev.), vol. 1, Pt. 2: 533-1327, Holmiae.

Lobanov, A. M., 1960: On the biology and ecology of Microchrysa polita L. (Diptera, Stratiomyiidae). - Ént. Obozr., 39: 341-348 (In Russian).

Loew, H., 1845: Dipterologische Beiträge. 52 pp., Posen.

- 1846a: Bemerkungen über die Gattung Beris und Beschreibung eines Zwitters von Beris nitens. - Stettin. ent. Ztg., 7: 219-224.

- 1846b: Fragmente zur Kenntnis der europäischen Arten einiger Dipterengattungen. - Linnaea ent., 1: 319-530.

- 1871: Beschreibung europäischer Dipteren. Systematische Beschreibung der bekannten europäischen zweiflügeligen Insecten, von Johann Wilhelm Meigen. Vol. 2: Neunter Teil oder dritter Supplementband. 319 pp., Halle.

Lundbeck, W., 1907: Diptera Danica. Pt. 1. - Stratiomyiidae, etc. Copenhagen.

- 1914: Nogle sjældnere samt nogle for vor Fauna ny Dipterer I. - Ent. Meddr, 10: 100-111.

- 1919: Nogle sjældnere samt nogle for vor Fauna ny Dipterer II. - Vidensk. Meddr dansk naturh. Foren., 70: 227-250.

Lundblad, O., 1950: Studier över Insektfaunan i Fiby urskog. - K. svenska Vetensk Akad. Avh. Naturskydd., 6, 235 pp.

- 1954: Studier över insektfaunan i Uppsala universitets naturpark vid Vårdsätra. - K. svenska VetenskAkad. Avh. Naturskydd., 8, 67 pp.

- 1955: Studier över insektfaunan i Harparbol lund. - K. svenska VetenskAkad. Avh. Naturskydd., 13, 132 pp.

Lyneborg, L., 1960: Tovinger II. Almindelig del. Våbenfluer, Klæger m. fl. - Danm. Fauna, 66, 233 pp., København.

- 1965: 9. Diptera, Brachycera & Cyclorrhapha - Fluer. In: Hansted-Reservatets Entomologi. The Entomology of the Hansted Reservation, North Jutland, Denmark. - Ent. Meddr, 30: 201-262.

Macquart, J., 1834: Histoire naturelle des Insectes. Diptères. Vol. 1. 578 pp., Paris.

McFadden, M. W., 1967: Soldier fly larvae in America north of Mexico. - Proc. U. S. natn. Mus., 121: 1-72.

- 1972: The Soldier Flies of Canada and Alaska (Diptera: Stratiomyidae) - I. Beridinae, Sarginae and Clitellariinae. - Can. Ent., 104: 531-562.

Meigen, J. W., 1804: Klassifikazion und Beschreibung der europäischen zweiflügeligen Insecten (Diptera Linn.). Erster Band. Abt. 1, 152 pp., Braunschweig.

- 1820-30: Systematische Beschreibung der bekannten europäischen zweiflügeligen Insekten. Vol. 2 (1820), 365 pp.; Vol. 3 (1822), 416 pp.; Vol. 6 (1830), 401 pp. Aachen, Hamm.

Meijere, J. C. H. de, 1916: Beiträge zur Kenntnis der Dipterenlarven und -puppen. - Zool. Jb., 40: 177-322.

Müller, G. W., 1925: Kalk in der Haut der Insekten und die Larve von Sargus cuprarius L. - Z. Morph. Ökol. Tiere, 3: 542-566.

Nagatomi, A., 1968: Exodontha dubia (Zetterstedt) new to Japan. - Kontyû, 36: 255-258.

Nagatomi, A. & Tanaka, A., 1971: The Solvidae of Japan. - Mushi, 45: 101-146.

Nartshuk, E. P., 1969: Stratiomyidae, in: Opredelitel nasekomych evropeiskoj časti SSSR. Vol. 5 (1): 454-481.

Panzer, G. W. F., 1798: Faunae insectorum germanicae initiae oder Deutschlands Insecten. H. 54, 24 pp., 24 pls. Nürnberg.

Papp, L., 1971: Ecological and production biological data on the significance of flies breeding in cattle droppings. - Acta zool. hung., 17: 91-105.

Pleske, T., 1924: Etude sur les Stratiomyiidae de la région paléarctique - II. Revue des especes paléarctique de la sous famille des Pachygasterinae. - Encycl. Ent. B II (Dipt.), 1: 95-103.

- 1926: Revision des especes paléarctiques des familles Erinnidae et Coenomyiidae. - Encycl. Ent. B II (Dipt.), 2 (1925): 161-184.

- 1930: Résultats scientifiques des Expéditions entomologiques du Musée Zoologique dans la région de l'Oussouri. - II. Diptera: Les Stratiomyiidae, Erinnidae, Coenomyiidae et Oestridae. - Ezheg. zool. Mus. Leningrad, 31: 181-206.

Ringdahl, O., 1914: Fyndorter för Diptera. - Ent. Tidskr., 35: 69-77.

- 1917: Fyndorter för Diptera. - Ent. Tidskr., 38: 302-311.

- 1931: Insektfaunan inom Abisko Nationalpark III. - 6. Flugor - Diptera Brachycera. - K. svenska VetenskAkad. Avh. Naturskydd., 18: 1-32.

- 1935: Fyndorter för sydsvenska Diptera. - Ent. Tidskr., 56: 201-204.

- 1937: Om djurvärlden i Råå kärr. - Skånes Natur, 1937: 77-83.

- 1939: Diptera Brachycera i Regio alpina. - Ent. Tidskr., 60: 37-50.

- 1941: Bidrag till kännedomen om flugfaunan (Diptera Brachycera) på Hallands Väderö. - Ent. Tidskr., 62: 1-23.

- 1947: Förteckning över flugor från Ölands alvar. - Ent. Tidskr., 68: 21-28.

- 1951: Flugor från Lapplands, Jämtlands och Härjedalens Fjälltrakter (Diptera Brachycera). - Opusc. ent., 16: 113-186.

- 1952: Småplock ur Kullabergs flugfauna. - Fauna och flora, 1952: 234-244.

- 1954: Några dipterologiska anteckningar från Råå kärr och vassar. - Ent. Tidskr., 75: 223-234.

- 1958: Dipterologiska notiser 16 och 17. - Opusc. ent., 23: 90-94.

- 1959a: Flugor på strandängar i nordvästra Skåne. - Ent. Tidskr., 80: 39-47.

- 1959b: Några tillägg till flugfaunan på Hallands Väderö. - Ent. Tidskr., 80: 120-122.

- 1960: Flugfaunan på Kullaberg och Hallands Väderö. - Kullabergs Natur, 2: 1-40.

Sahlberg, C. R., 1839: Några anmärkningar om Xylophagus maculatus. - Acta Soc. Sci. fenn., 1: 163-168.

Scopoli, J. A., 1763: Entomologia carniolica exhibens insecta carnioliae indigene et distributa in ordines, genera, species, varietates methodo Linnaeana. 421 pp., Vindobonae.

Schøyen, W. M., 1889: Supplement til H. Siebke's Enumeratio Insectorum Norvegicorum. Fasc. IV. (Diptera). 15 pp., Christiania.

Schremmer, F., 1951: Zur Biologie der Larve von Hermione (Oxycera) calceata und Hermione meigeni Staeg. (Diptera, Strat.). Zugleich ein Beitrag zur fauna hygropetrica. - Öst. zool. Z., 3: 126-139.

Séguy, E., 1926: Diptères (Brachycères). Stratiomyiidae, etc. - Faune Fr. Vol. 13, 308 pp., Paris.

Siebke, H., 1863: Beretning om en i Sommeren 1861 foretagen entomologisk Reise. - Nyt Mag. Naturvid., 12: 105-192.

- 1877: Enumeratio Insectorum Norvegicorum. Fasc. IV. - Catalogus Dipterorum Continentem. 255 pp., Christianiae.

Smith, K.G.V., 1957: Some miscellaneous records of bred Diptera. - Entomologist's Rec. J. Var., 69: 214-216.

Stackelberg, A.A., 1954: Materialy po faune dvukrylych Leningradskoj oblasti. - II. Diptera Brachycera. - Trudy Zool. Inst. AN SSSR, 15: 199-228.

Stæger, R.C., 1844: Bemerkungen über Musca hypoleon Lin. - Stettin. ent. Ztg., 5: 403-410.

Storm, V., 1907: Supplerende iagttagelser over Insecta Diptera ved Trondhjem. - K. norske Vidensk. Selsk. Skr., 5.

Strand, E., 1903: Norske lokaliteter for Diptera. - K. norske Vidensk. Selsk. Forh., 3: 1-11.

Strobl, P.G., 1909: Die Dipteren von Steiermark. - II. Nachtrag. - Mitt. naturw. Ver. Steierm., 46: 45-293.

Vaillant, F., 1951: Les larves d'Hermione. - Trav. Lab. Hydrobiol. Piscic. Univ. Grenoble, 1951: 23-38.

Verrall, G.H., 1909: Stratiomyiidae, etc. British Flies V. London.

Wahlberg, P.F., 1854: Bidrag till kännedomen om de nordiska Diptera. - Öfvers. K. VetenskAkad. Förh., 11: 211-216.

Wahlgren, E., 1907: Diptera 1. - Första Underordningen. Orthorapha. - Andra gruppen. Flugor. Brachycera. Fam. 14-23. - Svensk Insektfauna, 11, 1 (2), 62 pp., Uppsala.

- 1921: En för Sverige ny Xylomyia-art. - Ent. Tidskr., 42: 125.

Wallengren, H.D.J., 1866: Nordöstra Skånes Fauna. - Öfvers. K. VetenskAkad. Förh., 1866 (1): 3-15.

- 1870: Anteckningar i Entomologi. - Öfvers. K. VetenskAkad. Förh., 1870 (3): 145-182.

Zetterstedt, J.W., 1838: Insecta Lapponica. 1140 pp., Lipsiae.

- 1842-60: Diptera Scandinaviae. Disposita et descripta. Vol. 1-14, 6609 pp., Lundae.

Zimsen, E., 1964: The type material of J.C. Fabricius. 656 pp., Copenhagen.

Index

Page references are given to the main treatment of genera and species. Valid names are underlined.

Author's address:
Department of Biology of Animals and Man,
Faculty of Science J. E. Purkyně University,
Kotlářská 2,
Brno, Czechoslovakia.

Catalogue

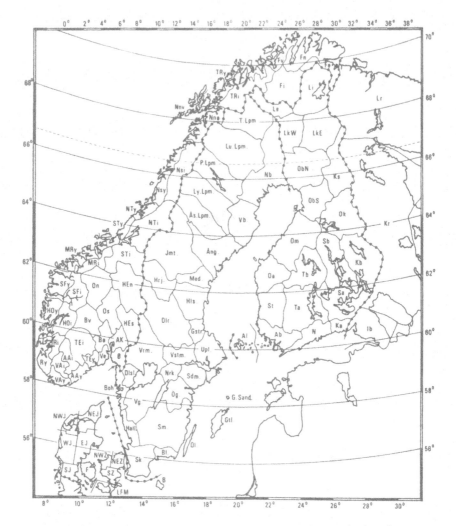

List of abbreviations for the provinces used throughout the text, on the map and in the following tables.

DENMARK

SJ	South Jutland	LFM	Lolland, Falster, Møn
EJ	East Jutland	SZ	South Zealand
WJ	West Jutland	NWZ	North West Zealand
NWJ	North West Jutland	NEZ	North East Zealand
NEJ	North East Jutland	B	Bornholm
F	Funen		

SWEDEN

Sk.	Skåne	Vrm.	Värmland
Bl.	Blekinge	Dlr.	Dalarna
Hall.	Halland	Gstr.	Gästrikland
Sm.	Småland	Hls.	Hälsingland
Öl.	Öland	Med.	Medelpad
Gtl.	Gotland	Hrj.	Härjedalen
G. Sand.	Gotska Sandön	Jmt.	Jämtland
Ög.	Östergötland	Ång.	Ångermanland
Vg.	Västergötland	Vb.	Västerbotten
Boh.	Bohuslän	Nb.	Norrbotten
Dlsl.	Dalsland	Ås. Lpm.	Åsele Lappmark
Nrk.	Närke	Ly. Lpm.	Lycksele Lappmark
Sdm.	Södermanland	P. Lpm.	Pite Lappmark
Upl.	Uppland	Lu. Lpm.	Lule Lappmark
Vstm.	Västmanland	T. Lpm.	Torne Lappmark

NORWAY

Ø	Østfold	HO	Hordaland
AK	Akershus	SF	Sogn og Fjordane
HE	Hedmark	MR	Møre og Romsdal
O	Opland	ST	Sør-Trøndelag
B	Buskerud	NT	Nord-Trøndelag
VE	Vestfold	Ns	southern Nordland
TE	Telemark	Nn	northern Nordland
AA	Aust-Agder	TR	Troms
VA	Vest-Agder	F	Finnmark
R	Rogaland		

n northern   s southern   ø eastern   v western   y outer   i inner

FINLAND

Al	Alandia	Kb	Karelia borealis
Ab	Regio aboensis	Om	Ostrobottnia media
N	Nylandia	Ok	Ostrobottnia kajanensis
Ka	Karelia australis	ObS	Ostrobottnia borealis, S part
St	Satakunta	ObN	Ostrobottnia borealis, N part
Ta	Tavastia australis	Ks	Kuusamo
Sa	Savonia australis	LkW	Lapponia kemensis, W part
Oa	Ostrobottnia australis	LkE	Lapponia kemensis, E part
Tb	Tavastia borealis	Li	Lapponia inarensis
Sb	Savonia borealis		

USSR

Ib   Ingria borealis   Kr   Karelia rossica   Lr   Lapponia rossica

		DENMARK												
		N. Germany / G. Britain	SJ	EJ	WJ	NWJ	NEJ	F	LFM	SZ	NWZ	NEZ	B	Sk. / Bl.

Family SOLVIDAE

Species	No.	N. Germany	G. Britain	SJ	EJ	WJ	NWJ	NEJ	F	LFM	SZ	NWZ	NEZ	B	Sk.	Bl.
Solva interrupta Pleske	1															
S. maculata (Meig.)	2		●						●				●		●	
S. marginata (Meig.)	3	●	●										●		●	

Family STRATIOMYIDAE

Species	No.	N. Germany	G. Britain	SJ	EJ	WJ	NWJ	NEJ	F	LFM	SZ	NWZ	NEZ	B	Sk.	Bl.
Beris chalybeata (Forst.)	1	●	●	●	●		●	●	●	●			●		●	●
B. clavipes (L.)	2	●	●	●	●				●				●		●	
B. fuscipes Meig.	3	●	●													
B. morrisii Dale	4	●	●	●	●				●	●			●		●	
B. strobli Duš. & Rozk.	5	●														
B. vallata (Forst.)	6	●	●	●	●	●			●	●	●	●	●	●	●	
Exodontha dubia (Zett.)	7															
Microchrysa cyaneiventris (Zett.)	8		●	●	●								●			●
M. flavicornis (Meig.)	9	●	●	●	●	●	●	●	●	●	●		●	●	●	
M. polita (L.)	10	●	●	●	●	●	●	●	●	●	●		●	●	●	
Sargus cuprarius (L.)	11	●	●	●	●	●	●	●	●	●	●		●	●	●	
S. iridatus (Scop.)	12	●	●	●	●	●	●	●	●	●	●	●	●	●	●	●
S. rufipes Wahlbg.	13															
S. splendens Meig.	14	●	●	●	●		●	●	●		●	●		●	●	●
Chloromyia formosa (Scop.)	15	●	●	●	●		●	●	●	●	●	●	●	●	●	
Stratiomys chamaeleon (L.)	16	●	●	●	●			●			●			●		
S. furcata Fabr.	17	●	●	●			●	●	●	●	●	●	●	●		
S. longicornis (Scop.)	18	●	●	●	●			●	●	●	●		●			
S. potamida Meig.	19	●	●	●	●	●				●			●	●		
Odontomyia angulata (Panz.)	20	●	●		●				●	●	●		●			
O. argentata (Fabr.)	21	●	●							●	●		●			
O. hydroleon (L.)	22	●	●		●				●	●	●		●			
O. microleon (L.)	23	●			●		●	●					●			
O. ornata (Meig.)	24	●	●	●	●				●	●			●			●
O. tigrina (Fabr.)	25	●	●	●	●		●	●	●	●	●	●	●	●		
Oplodontha viridula (Fabr.)	26	●	●	●	●	●	●	●	●	●	●		●	●		
Nemotelus nigrinus Fall.	27	●	●	●				●	●	●	●		●			
N. notatus Zett.	28	●	●	●	●	●	●	●	●	●	●	●	●	●	●	
N. pantherinus (L.)	29	●	●	●	●				●	●		●	●	●		
N. uliginosus (L.)	30	●	●	●	●	●		●	●	●	●		●			
Clitellaria ephippium (Fabr.)	31	●	●													

SWEDEN

No.	Hall.	Sm.	Öl.	Gtl.	G. Sand.	Ög.	Vg.	Boh.	Dlsl.	Nrk.	Sdm.	Upl.	Vstm.	Vrm.	Dlr.	Gstr.	Hls.	Med.	Hrj.	Jmt.	Äng.	Vb.	Nb.	Ås. Lpm.	Ly. Lpm.	P. Lpm.	Lu. Lpm.	T. Lpm.
1																												
2																												
3																												
1		●					●	●							●	●		●		●	●			●	●			
2	●	●	●	●		●	●	●		●		●		●	●		●					●	●					
3																									●		●	
4		●				●																						
5												●								●	●	●					●	●
6			●																									
7																				●					●	●		
8		●	●									●			●													
9	●	●	●	●	●	●	●					●								●								
10	●	●	●	●	●	●	●	●	●	●	●			●	●					●	●	●					●	●
11	●	●	●	●	●	●	●							●	●					●	●	●				●		
12	●	●	●	●	●	●	●							●	●					●	●	●	●				●	●
13															●					●	●						●	●
14		●	●	●	●		●			●		●			●	●		●		●		●						
15	●	●	●	●		●	●																					
16	●			●	●						●	●								●								
17	●		●	●		●	●						●	●								●						
18																												
19																												
20		●	●	●								●																
21				●	●							●	●															
22				●	●							●								●								
23		●		●	●							●																
24				●																								
25		●		●	●					●		●																
26	●	●		●	●							●								●								
27		●		●		●						●								●		●		●	●			●
28		●	●			●	●																					
29		●	●	●							●																	
30	●	●	●			●	●	●		●																		
31		●	●																									

		Ø+Ak	HE (s+n)	O (s+n)	B (ø+v)	VE	TE (y+i)	AA (y+i)	VA (y+i)	R (y+i)	HO (y+i)	SF (y+i)	MR (y+i)	ST (y+i)	NT (y+i)	Ns (y+i)
Family SOLVIDAE																
Solva interrupta Pleske	1															
S. maculata (Meig.)	2															
S. marginata (Meig.)	3															
Family STRATIOMYIDAE																
Beris chalybeata (Forst.)	1	●									●			●		
B. clavipes (L.)	2	●	●							●	●			●	●	
B. fuscipes Meig.	3															
B. morrisii Dale	4															
B. strobli Duš. & Rozk.	5															
B. vallata (Forst.)	6															
Exodontha dubia (Zett.)	7	●	●								●					●
Microchrysa cyaneiventris (Zett.)	8									●						
M. flavicornis (Meig.)	9	●	●							●	●				●	
M. polita (L.)	10	●	●		●					●	●				●	
Sargus cuprarius (L.)	11	●									●			●		
S. iridatus (Scop.)	12	●	●		●	●					●				●	
S. rufipes Wahlbg.	13		●													
S. splendens Meig.	14	●	●							●	●			●		
Chloromyia formosa (Scop.)	15	●								●	●					
Stratiomys chamaeleon (L.)	16															
S. furcata Fabr.	17	●														
S. longicornis (Scop.)	18															
S. potamida Meig.	19															
Odontomyia angulata (Panz.)	20															
O. argentata (Fabr.)	21															
O. hydroleon (L.)	22	●														
O. microleon (L.)	23	●		●												
O. ornata (Meig.)	24															
O. tigrina (Fabr.)	25															
Oplodontha viridula (Fabr.)	26	●					●							●		
Nemotelus nigrinus Fall.	27	●	●											●	●	
N. notatus Zett.	28	●														
N. pantherinus (L.)	29		●													
N. uliginosus (L.)	30	●			●									●		
Clitellaria ephippium (Fabr.)	31	●														

	Nn (ø+v)	TR (y+i)	F (v+i)	F (n+ø)	Al	Ab	N	Ka	St	Ta	Sa	Oa	Tb	Sb	Kb	Om	Ok	ObS	ObN	Ks	LkW	LkE	Lc	Li	Ib	Kr	Lr
1									●	●															●	●	
2																											
3																											
1			◐			●	●		●	●	●		●												●	●	
2	◐				●	●	●																				
3	◐	◐											●													●	
4						●	●		●	●	●						●		●						●	●	
5						●	●		●	●			●	●					●	●							
6																											
7																											
8					●	●	●		●	●	●		●	●				●							●	●	
9					●	●	●		●	●	●	●	●	●	●		●		●						●	●	
10	◐				●	●	●		●	●	●	●	●	●	●	●	●		●						●	●	
11		◐			●	●	●		●	●	●	●	●	●	●	●	●	●	●						●	●	
12	◐				●	●	●		●	●	●	●	●	●	●	●	●	●	●				●		●	●	●
13		◐																	●				●			●	
14					●	●	●	●		●		●							●								
15					●																						
16																										●	
17					●	●	●		●	●			●				●								●	●	
18																											
19																											
20					●																						
21						●																					
22					●																					●	
23					●	●	●		●	●				●	●		●								●	●	
24																											
25																											
26					●	●	●		●	●	●	●		●	●	●	●		●						●	●	
27	◐				●	●	●		●	●				●			●		●		●		●		●	●	
28					●	●																					
29																										●	
30					●	●	●				●		●	●	●	●		●							●	●	
31																											

Family STRATIOMYIDAE		N. Germany	G. Britain	SJ	EJ	WJ	NWJ	NEJ	F	LFM	SZ	NWZ	NEZ	B	Sk.	Bl.
Oxycera dives Loew	32		●													
O. fallenii Stæg.	33	●	●		●								●		●	
O. formosa Meig.	34	●	●		●				●	●			●		●	
O. freyi Lind.	35															
O. leonina (Panz.)	36	●											●			
O. meigenii Stæg.	37	●							●	●			●		●	
O. pardalina Meig.	38	●	●													
O. pygmaea (Fall.)	39	●	●								●	●	●		●	
O. terminata Meig.	40	●	●										●			
O. trilineata (L.)	41	●	●	●	●				●	●	●		●	●	●	
Pachygaster atra (Panz.)	42	●	●												●	
Praomyia leachii (Curt.)	43	●	●													
Neopachygaster meromelaena (Duf.)	44	●	●													
Eupachygaster tarsalis (Zett.)	45		●							●			●			
Zabrachia minutissima (Zett.)	46	●	●						●	●			●			
Berkshiria albistylum Johns.	47															

DENMARK

SWEDEN

	Hall.	Sm.	Öl.	Gtl.	G. Sand.	Ög.	Vg.	Boh.	Dlsl.	Nrk.	Sdm.	Upl.	Vstm.	Vrm.	Dlr.	Gstr.	Hls.	Med.	Hrj.	Jmt.	Äng.	Vb.	Nb.	Ås. Lpm.	Ly. Lpm.	P. Lpm.	Lu. Lpm.	T. Lpm.
32																												
33																												
34																												
35																									●			
36																												
37																												
38		●					●																					
39		●		●			●													●								
40																												
41		●	●	●	●	●	●	●																				
42				●			●																					
43			●	●																								
44				●		●						●																
45				●																								
46		●				●	●					●		●						●								
47		●								●		●																

Family STRATIOMYIDAE		Ø+Ak	HE (s+n)	O (s+n)	B (ø+v)	VE	TE (y+i)	AA (y+i)	VA (y+i)	R (y+i)	HO (y+i)	SF (y+i)	MR (y+i)	ST (y+i)	NT (y+i)	Ns (y+i)
Oxycera dives Loew	32															
O. fallenii Stæg.	33															
O. formosa Meig.	34															
O. freyi Lind.	35															
O. leonina (Panz.)	36															
O. meigenii Stæg.	37															
O. pardalina Meig.	38															
O. pygmaea (Fall.)	39															
O. terminata Meig.	40															
O. trilineata (L.)	41	▶														
Pachygaster atra (Panz.)	42															
Praomyia leachii (Curt.)	43															
Neopachygaster meromelaena (Duf.)	44															
Eupachygaster tarsalis (Zett.)	45															
Zabrachia minutissima (Zett.)	46															
Berkshiria albistylum Johns.	47															

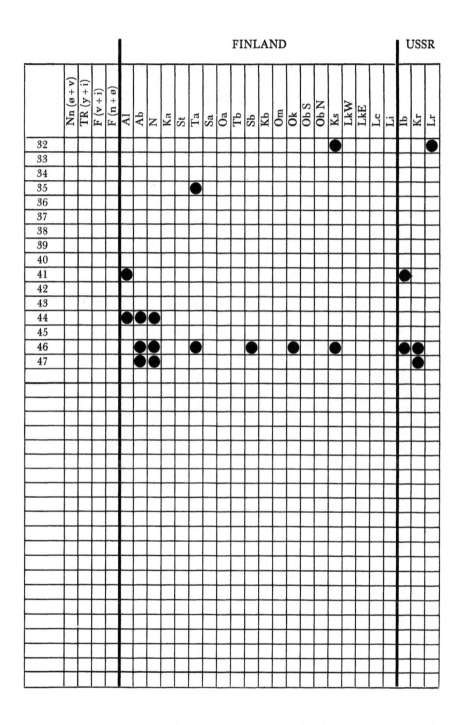